SpringerBriefs in Education

We are delighted to announce SpringerBriefs in Education, an innovative product type that combines elements of both journals and books. Briefs present concise summaries of cutting-edge research and practical applications in education. Featuring compact volumes of 50 to 125 pages, the SpringerBriefs in Education allow authors to present their ideas and readers to absorb them with a minimal time investment. Briefs are published as part of Springer's eBook Collection. In addition, Briefs are available for individual print and electronic purchase.

SpringerBriefs in Education cover a broad range of educational fields such as: Science Education, Higher Education, Educational Psychology, Assessment & Evaluation, Language Education, Mathematics Education, Educational Technology, Medical Education and Educational Policy.

SpringerBriefs typically offer an outlet for:

- An introduction to a (sub)field in education summarizing and giving an over-view of theories, issues, core concepts and/or key literature in a particular field
- A timely report of state-of-the art analytical techniques and instruments in the field of educational research
- A presentation of core educational concepts
- An overview of a testing and evaluation method
- A snapshot of a hot or emerging topic or policy change
- An in-depth case study
- A literature review
- A report/review study of a survey
- An elaborated thesis

Both solicited and unsolicited manuscripts are considered for publication in the SpringerBriefs in Education series. Potential authors are warmly invited to complete and submit the Briefs Author Proposal form. All projects will be submitted to editorial review by editorial advisors.

SpringerBriefs are characterized by expedited production schedules with the aim for publication 8 to 12 weeks after acceptance and fast, global electronic dissemination through our online platform SpringerLink. The standard concise author contracts guarantee that:

- an individual ISBN is assigned to each manuscript
- each manuscript is copyrighted in the name of the author
- the author retains the right to post the pre-publication version on his/her website or that of his/her institution

More information about this series at http://www.springer.com/series/8914

Orit Hazzan · Einat Heyd-Metzuyanim
Anat Even-Zahav · Tali Tal
Yehudit Judy Dori

Application of Management
Theories for STEM
Education

The Case of SWOT Analysis

Springer

Orit Hazzan
Technion – Israel Institute of Technology
Haifa
Israel

and

The Samuel Neaman Institute for National
Policy Research Technion City
Haifa
Israel

Einat Heyd-Metzuyanim
Technion – Israel Institute of Technology
Haifa
Israel

Anat Even-Zahav
Technion – Israel Institute of Technology
Haifa
Israel

Tali Tal
Technion – Israel Institute of Technology
Haifa
Israel

and

The Samuel Neaman Institute for National
Policy Research Technion City
Haifa
Israel

Yehudit Judy Dori
Technion – Israel Institute of Technology
Haifa
Israel

and

The Samuel Neaman Institute for National
Policy Research Technion City
Haifa
Israel

ISSN 2211-1921   ISSN 2211-193X (electronic)
SpringerBriefs in Education
ISBN 978-3-319-68949-4   ISBN 978-3-319-68950-0 (eBook)
https://doi.org/10.1007/978-3-319-68950-0

Library of Congress Control Number: 2017954486

Printed on acid-free paper

This Springer imprint is published by Springer Nature
The registered company is Springer International Publishing AG
The registered company address is: Gewerbestrasse 11, 6330 Cham, Switzerland

Preface

The Brief comprises of three chapters that describe the application of management theories in Science, Technology, Engineering and Mathematics (STEM) education systems: Two chapters examine STEM education on the K-12 national level (Chaps. 1 and 3), and one chapter focuses on the higher education institutional level (Chap. 2). All chapters are based on comprehensive research works. Thus, it might appeal to all practitioners who care about STEM education: in the schools, the academia, and the government, who hold a wide variety of roles such as teachers, school principals, researchers, graduate students, and policy makers from the government. In all the three chapters, we use Strengths, Weaknesses, Opportunities, and Threats (SWOT) analysis as a managerial strategic tool for the examination of factors that focus either on internal circumstance—strengths and weaknesses—or on external ones—opportunities and threats.

Our illustration of SWOT analysis of educational organizations is relevant due to the increased awareness of the similarities existing between educational organizations and other kinds of (both for-profit and nonprofit) organizations. Accordingly, it has been used also as a tool for the analysis of public organizations, such as schools and hospitals. When needed, we elaborate on the SWOT analysis framework throughout the Brief.

Specifically, in Chap. 1, the application of SWOT analysis is illustrated in the case of STEM education on the national level. Chapter 2 demonstrates the application of SWOT analysis in the academia. Chapter 3 uses SWOT analysis for the examination of Research–Practice Partnerships, which bridge the gap between the education system and the academia.

In Chap. 1—*STEM Teachers' SWOT Analysis of STEM Education: The Bureaucratic–Professional Conflict*—Even-Zahav and Hazzan characterize the way in which STEM teachers perceive (a) their profession, (b) the education system in which they teach, and (c) the way in which the system perceives them and the profession of STEM teaching. Data analysis revealed five conflicts that STEM teachers deal with, which focus on the teachers as professionals in the organization for which they work: (a) the professional opportunities conflict; (b) the profession perception conflict; (c) the discourse on STEM education conflict; (d) the academic freedom conflict; and (e) the teacher's status conflict.

In Chap. 2—*SWOT Analysis of STEM Education in Academia: The Disciplinary versus Cross Disciplinary Conflict*—Dori, Tal, and Heyd-Metzuyanim show how to turn a SWOT analysis process of an academic department or faculty into a strategic plan in a way that fosters the strengths, eliminates the weaknesses, exhausts opportunities, and deals with outside threats. The case described in this chapter focuses on the Technion's Faculty of Education in Science and Technology. The SWOT characteristics, though specific to this particular Faculty, result from the tensions between disciplinary and interdisciplinary teaching and research. These tensions can be observed in content knowledge, pedagogical principles and approaches, as well as research methods. Besides these tensions, there are also constraints of funding, human resources, and workload of both faculty and students, which are common to all disciplines in higher education. As such, the analysis can serve as a useful managerial tool for coping with challenges that a STEM faculty or a dean face in the process of searching for creative avenues for growth and improvement in their teaching and research.

In Chap. 3—*Research–Practice Partnerships in STEM Education: An Organizational Perspective*—Hazzan, Heyd-Metzuyanim, and Dori examine Research–Practice Partnerships (RPPs) in Israeli STEM education. Based on SWOT analysis, relevant factors were categorized into strengths, weaknesses, opportunities, and threats related to the scene of STEM education in Israel. The SWOT analysis presented in this chapter, which is based on data gathered from representatives of all STEM education sectors, enables to deepen the understanding of how RPPs in STEM education can be enhanced and improved.

As can be seen, all chapters suggest useful guidelines for STEM education organizations, either on the national level or on the higher education institutional level, for policy makers, who wish to improve the competency of their organization, department, or group, in both teaching and research. Such improvement aligns with the new era and directions expected in STEM education.

Haifa, Israel   Orit Hazzan
                                                          Einat Heyd-Metzuyanim
                                                                Anat Even-Zahav
                                                                          Tali Tal
                                                              Yehudit Judy Dori

Contents

Chapter 1
STEM Teachers' SWOT Analysis of STEM Education: The Bureaucratic–Professional Conflict

Anat Even-Zahav & Orit Hazzan

Abstract This chapter describes a study that aims to characterize the perception of STEM teachers of the Israeli STEM education system. It analyzes the bureaucratic–professional conflict that the teachers experience which addresses the conflict between teachers' professional norms and the organizational norms. The main tool of the research was in-depth interviews with STEM teachers, designed according to the four components of the SWOT analysis with respect to STEM education in Israel. The analysis indicated that STEM teachers' perceptions of the STEM education system in Israel can be characterized by five conflicts the teachers face. The conflicts express the gap between the teachers' perceptions and the organizational perception of STEM education in Israel. The five conflicts are the *professional opportunities* conflict, the *profession perception* conflict, the *discourse on STEM education* conflict, the *academic freedom* conflict, and the *teacher status* conflict. The conflicts suggest the need to change the structure of STEM education in Israel as well as the need for a proactive approach in positioning and promoting STEM education.

Keywords STEM teachers · STEM education · Bureaucratic–professional conflict · SWOT analysis

1.1 Introduction

This article describes a study that aims to characterize the way in which STEM teachers[1] perceive (a) their profession, (b) the education system in which they teach, and (c) the way in which the system perceives them and the profession of STEM teaching.

---

[1]STEM—Science, Technology, Engineering and Mathematics; In this study, "STEM teachers" refer to teachers who teach the following five subjects: mathematics, physics, chemistry, biology, and computer science.

© The Author(s) 2018
O. Hazzan et al., *Application of Management Theories for STEM Education*,
SpringerBriefs in Education, https://doi.org/10.1007/978-3-319-68950-0_1

In the Global Competitiveness Report 2012–2013 of the World Economic Forum, published in September 2012, Israel ranked third in the world for its innovative ability, but only 89th for the achievements of its students in mathematics and science (World Economic Forum 2012). In light of these findings, it is necessary to improve STEM education in Israel. Indeed, several reports have been written over the past few years, in Israel in particular and around the world in general, that address STEM education (Initiative for Applied Education Research 2012; PCAST 2010). However, an examination of these reports reveals **the absence of the teachers' voice**—those men and women of the profession, who are well familiar with its difficulties and challenges. The objective of the study was, therefore, to characterize STEM teachers' perception of STEM education in Israel's educational system. This objective led to the research question: How do STEM teachers perceive STEM education in Israel?

The primary research tool used to collect data was in-depth interviews with 12 senior STEM teachers (hereinafter the "STEM teachers") from high schools throughout Israel. In the interviews, the teachers were asked to perform a SWOT analysis of the Israeli educational system. SWOT (Strengths, Weaknesses, Opportunities, Threats) analysis is a method used to examine the correspondence of an organization's strategy to the environment in which it operates. Data were collected also by an online questionnaire sent to the participating schools.

Data analysis revealed five different conflicts the STEM teachers deal with, which focus on the teachers as professionals in the organization for which they work: (a) the professional opportunities conflict; (b) the profession perception conflict; (c) the discourse on STEM education conflict; (d) the academic freedom conflict; and (e) the teacher's status conflict.

In what follows, the Theoretical Background reviews the essence of the bureaucratic–professional conflict, the features of professionalism, sources of the development of the conflict, as well as careers and professional development in educational organizations. The research structure section presents the research tools used in the study and the research participants. The Findings section describes the five different manifestations of the bureaucratic–professional conflict. In the Summary, we shall position these conflicts within the discourse on STEM education in Israel.

1.2 Theoretical Background

1.2.1 The Bureaucratic–Professional Conflict

The organizational sociologist Afzalur Rahim (2002) provides the following definition of conflict:

> Conflict as an interactive process manifested in incompatibility, disagreement, or dissonance within or between social entities (i.e., individual, group, organization, etc.). (p. 207)

Specifically, a **bureaucratic–professional conflict** arises where professional norms and organizational or bureaucratic norms clash, and the conflict does not enable the professionals to realize their objectives (Hall 1967).

Conflict management is the implementation of strategies whose purpose is to limit the negative aspects of a conflict while enhancing its positive aspects, to the point where the positive aspects either equal or surpass the negative ones. Furthermore, the aim of conflict management is the improvement of learning, which in turn leads to enhanced effectiveness and performance of the organization. The goal of conflict management is not to make all conflicts disappear or to avoid conflicts, since conflicts may at times be of value to groups and organizations (Afzalur Rahim 2002).

In schools, the bureaucratic–hierarchical structure emphasizes authority in which the teachers must obey the principal. The decisions of the school principal, the bureaucrat, are guided in many cases by the policy of the organization, i.e., the education system, whereas those of the teachers, the professional ones, are guided by their professional knowledge and accumulated experience; thus, a conflict is created (Oplatka 2007).

The STEM teachers who participated in this study were perceived as professionals, while at the same time, exhibited a conflict as expressed in the study findings (see Sect. 1.4).

1.2.2   Characteristics of Professionalism

In order to understand the professional norms according to which professionals operate, the characteristics of professional behavior are presented. Hall (1968) defined a five-characteristic scale to classify professionalism. This scale later served as a basis for comparison with alternative scales used to evaluate professionalism in a variety of professional groups (Shafer et al. 2002). The STEM teachers who participated in the study meet many of the criteria of these characteristics of professionalism, as the findings will demonstrate (see Sect. 1.4).

The five professional characteristics formulated by Hall (1968) which refer to professionalism in its broader context are described below, in addition to a discussion of professionalism by Hativa (2008), in the context of the teaching profession.

(*a*) *Professional community affiliation*: Professional community affiliation refers to the extent to which a person is actively involved in the professional community, utilizing professional institutions as a main referent. People remain closely identified with their professional community through activities such as reading professional publications and participating in conferences. According to Hativa (2008), who referred to the characteristics of professionalism as associated with teaching, teachers acquire over the years local knowledge only, based on their experience in a limited number of classes and schools. Nevertheless, the existence of a professional community of educators is vital, as it is through this community that its members can share their knowledge with one another through personal meetings, visits to classes, conferences, seminars and advanced education courses, and publications in professional journals.

(*b*) *Social obligation*: Recognition of social obligation, or the obligation to serve the public interest, is one of the basic characteristics of traditional professions. Professionals are expected to subordinate their own personal gain to serving the public, when the two come into conflict (Wilensky 1964). This approach led professions such as law and medicine to recognize an obligation to provide their services voluntarily or free of charge to those who need them, as part of their professional codes of behavior. In this context, teaching in public schools provides a vital service to society in that it trains and prepares youngsters for adult life (Hativa 2008).

(*c*) *Belief in self-regulation or regulation by the professional community*: According to traditional professional theories, a profession is granted monopoly rights to provide services by means of a license governed by law. In return, those practicing the profession must provide a high level of service. The rationale underlying the belief in self-regulation is that laymen are not qualified to judge the quality of a professional's work; hence, those most qualified to judge the work of professionals are their colleagues. In Israel, although the teaching profession lacks a professional community that regulates the quality of teaching and preserves teaching knowledge (Hativa 2008), curriculum is supervised and teaching standards are maintained for the different subjects through main supervisors who are the education system representatives for each particular subject.

(*d*) *Professional dedication*: Professional dedication reflects the sense of a "call to the field"—the faith that the professional wishes to perform the job, even when external remuneration is lacking. This is a dominant characteristic in the teaching profession. The role of a teacher in society is that of a social calling, since financial remuneration in this profession is not high, compared with several other professions that provide service to society. The OECD average (in 2013) indicates that teachers' salaries comprise 83% of the wages of academics with similar education (OECD 2015).

(*e*) *Autonomy demands*: The demand for autonomy refers to professionals' aspiration for freedom in the making of decisions regarding their work. External pressures that clash with professional discretion are the antithesis of such autonomy. Characteristics such as professional certification and the need for self-regulation encourage the demand for autonomy. With regard to teaching, professional teachers make professional judgments and decisions based on theoretical and practical knowledge on a daily basis, during class, while preparing a lesson or while communicating with colleagues or with the school administration (Hativa 2008).

1.2.3 Career and Professional Development

The findings of this study attest to a lack of professional opportunities for teachers, to a need for professional development, to a conflict regarding the teacher's status, and to job burnout among teachers. To understand how this conflict arises, a brief theoretical overview is presented, including the concept of career in an organization in general and in an educational organization in particular.

1.2.3.1 Career in an Organization

According to Hall (2002), a career is a way in which the individual perceives the continuum of positions and behaviors related to professional experiences and actions over the course of his or her work life. This definition emphasizes a subjective career—the manner in which a person perceives his or her career. The experiences consist of objective events such as a job change and of a subjective interpretation such as professional aspirations, expectations, and attitudes towards the job.

Oplatka (2007) notes that career management is important for both individuals and organizations, as it bridges the gap between the needs of the organization and those of its employees. Organizations rely upon people to execute their functions, while people rely on organizations to provide them with jobs and professional opportunities in their careers. Career management gains central importance and influences the adaption and development of the individual (Hall 2002).

The dual development ladder in organizations is one alternative for organizations to provide a career development offering two distinct development tracks: a professional track and a management track, with the possibility of transitioning from one to the other according to the employees' abilities and organizational needs. In their initial stage of development, organization employees follow the professional track. Later on, after accumulating job seniority, the employee and the organization decide which track to pursue. Thus, employees need not be obligated to transition to supervisory or management positions in order to advance in the organizational hierarchy (Cesare and Thornton 1993; Noe 2002).

1.2.3.2 Career in Educational Organizations

When formulating the school's educational–managerial policy, it is important to consider issues related to career development such as salary, promotion, and professional opportunities for teachers.

Barak (2011) presents the concept of a professional career ladder for teachers, the objective of which is to offer teachers the prospect of development and change, to attract people to the teaching profession who appreciate opportunities for professional and personal development, and to encourage professional growth.

According to Barak, this idea is now commonly accepted in many educational systems around the world that define stages for promotion. Promotion is defined as the assumption of responsibility and positions of leadership in a school, a district, or nationwide. Examples of leadership positions include teachers responsible for professional development, teacher-coaches in various content areas, team-leading teachers, and teacher advisors. This system includes peer evaluation, which enables to identify the most effective teachers and train them for positions of leadership. Examples of career ladders in educational systems around the world are reviewed in Appendix 2.

In Israel, professional development frameworks are offered to teachers in secondary schools as part of the "Courage for Change"[2] reform (implemented since 2011). The professional development frameworks offered in this reform are as follows:

(1) school-based professional development that focuses on improving teaching and learning and on deepening the dialog between teachers and students, and between teachers and the general school population;
(2) professional development for position-holders, i.e., preparatory training for jobs such as subject leader, class coordinator, homeroom teacher, including short courses at institutions that are recognized by the Ministry of Education (universities and academic colleges);
(3) professional development for teachers in general education disciplines and subjects, as formulated by the professional departments of the Ministry of Education, which may include didactic–educational and administrative–organizational issues.

This section focused on three concepts and their interpretation in the context of the Israeli education system: the bureaucratic–professional conflict; the characteristics of professionalism; and teachers' careers and professional development. These concepts form the basis for answering the research question: How do STEM teachers perceive STEM education in Israel?

1.3 Research Tools and Participants

This section describes the research tools used in this study and their implementation: the SWOT interview, observations in schools, and an online questionnaire.

---

[2]Reform "courage for change" (in Hebrew, Oz Latmura) is an agreement signed between teachers' unions and the education system in Israel in 2012 and aims to bring about a significant change in the top division (high school) in pedagogical aspects, administrative aspects and conditions of teachers' employment.

Strengths	Weaknesses
Individual level	Individual level
School level	School level
Curriculum level	Curriculum level
Education system level	Education system level
Opportunities	**Threats**
Individual level	Individual level
School level	School level
Curriculum level	Curriculum level
Education system level	Education system level

Fig. 1.1 SWOT interview structure

1.3.1 SWOT Interview

In our attempt to characterize the teachers' perception of STEM education in Israel, we constructed an interview along the lines of the SWOT analysis, which enables an analysis of organizations according to their strengths, their weaknesses, the opportunities that are open to them, and the threats they face. The interviews were conducted according to the four SWOT dimensions as they apply to four different levels: the individual, the school, the curriculum, and the education system (see Fig. 1.1). Thus, each interview in essence constituted both a SWOT analysis of the teacher and a SWOT analysis of the system, on three levels: the school, the curriculum, and the education system.

The SWOT interview was built as a semi-structured, in-depth interview and constituted the main source of information for this study. The in-depth interview suited the research question, which focused on understanding the STEM teachers' perception of the system and their experience of it, as well as the meaning they attribute to this experience (Patton 1990). We decided to shape the interview according to the SWOT components to direct the teachers to wear analysis lens at the education system, which examine it from the outside. This perspective differs from the more traditional perspective adopted in such interviews with practitioners from the organization.

The SWOT interviews were conducted with 12 teachers of the following five STEM subjects: mathematics, physics, chemistry, biology, and computer science.[3] These subjects were chosen to be focused on in the study since they are core subjects[4] (except for computer science).

The 12 participating teachers are seasoned teachers, with at least 10 years of teaching experience. They have experience teaching at the honor classes and

---

[3]Four mathematics teachers, 3 physics teachers, 2 chemistry teachers, 2 biology teachers, and 2 computer science teachers (see Appendix 1). One teacher teaches two subjects, hence, the sum is 13 (while the number of teachers is 12).

[4]Core subjects in Israel are mathematics, English, Bible, literature, chemistry, physics, biology, and history. Core subjects with a high-level matriculation examination are given a bonus factor when enrolling in university in Israel.

advanced placement courses (including preparing students for the highest level matriculation examinations). The teachers teach in a variety of junior high schools and high schools throughout Israel and in a variety of social sectors: state, state-religious, or charter schools. Most of the teachers hold some managerial or leadership role in addition to their teaching position such as department heads in their schools (see Appendix 1, STEM teachers—Academic and Professional Background). These facts attest to the seniority and professionalism of the teachers chosen for the study.

The SWOT interviews were analyzed in two stages. First, individual SWOT tables were constructed for each teacher, on the four levels: individual, school, curriculum, and education system; these tables were later consolidated into one table with 16 cells: four SWOT components multiplied by four levels for each component. Second, the analysis was performed in two cross sections: (a) analysis of connections between the various levels under each SWOT component and (b) analysis of connections between the four different SWOT components at the same level.

1.3.2   Observations in Schools

As part of the study, non-participant observations were conducted in each school prior to the interviews set in the school. In this observation, the researchers were external to the studied environment and not personally involved in it. The teachers were observed during random conversations with colleagues before and after the interviews, in the teachers' lounges and in offices of position-holders (deputy principal, pedagogical coordinator) as well as in some of the administration board meetings. In these observations, the STEM teachers could be seen voicing their opinions to other teachers and to school administration staff (mainly deputy principals) on issues related to the teaching of science in their school.

1.3.3   Online Survey

During the data analysis, the question arose whether or not STEM teachers tend to refrain from serving in administrative positions in their schools. To examine this question, we decided to administer an online survey whose purpose was to determine the proportion of STEM teachers (out of the overall teaching staff) serving in the following positions: class (year) coordinator, pedagogical coordinator, deputy principal, principal, or other administrative positions. Indeed, we found that in the schools participating in the study, the percentage of STEM teachers in administrative positions (mentioned above) was low. Data for these schools are presented in Appendix 3.

Data analysis led to the formulation of an organizing framework for the research findings, according to which the teachers' perceptions are described as conflicts that are manifested on three levels: the education system, the school, and the student.

1.4   Findings—The Expression of the Bureaucratic–Professional Conflict

This study reveals five conflicts the STEM teachers faced between STEM teachers' professional norms, on the one hand, and the organizational norms, on the other hand. These conflicts are organized and presented here (in most cases) on three levels: the education system, the school, and the students. Table 1.1 summarizes the conflicts on their respective levels.

Table 1.1 Five conflicts on the three levels[a]

Name of conflict	Level of conflict expression		
	Education system (1)	School (2)	Students (3)
(A) Professional opportunities	(A1) (a) Limited opportunities for career advancement (b) The need for a cooperative professional community	(A2) The need for professional development and for a cooperative professional community	(A3) Professional burnout and pedagogical difficulties
(B) Profession perception	(B1) Structure and content of STEM education in junior high schools	(B2) Loss of teaching hours due to school activities	(B3) (a) Student obligations in school (b) Dealing with students' discipline problems and personal matters
(C) STEM education discourse	(C1) The importance attributed to the percentage of matriculation eligibility versus the composition of the matriculation certificate	(C2) Ambiguous message regarding the importance of STEM education	(C3) The need to "market" the subject to students
(D) Academic freedom	(D1) Lack of autonomy regarding curriculum issues		
(E) Teacher's status	(E1) The teacher's perception of self versus the organization's regard for the teaching profession	(E2) Work conditions	

[a]Letters A to E signify the five conflicts (in addition to their names), and the numbers attached to the letters indicate the three levels: (1) the education system level, (2) the school level, and (3) the student level

Data analysis reveals that the bureaucratic–professional conflict is not the only conflict, but rather the main one of an entire series of conflicts. Furthermore, other conflicts were found that stem from teachers' expectations and do not necessarily result from their professional values, but rather from their expectations as employees in an organization (career advancement, teachers' status, etc.). The entire set of conflicts found in our research is presented in Table 1.1.

1.4.1   The Professional Opportunities Conflict (A)

One of the main findings is the gap exists between the STEM teachers' perceptions and the perceptions of the organization with respect to their professional development. This perception gap leads to a clash between the teacher and the organization to which he or she belongs, which is manifested on three levels: the education system level, the school level, and the student level. As we shall see, the clash at the student level stems from the clash at the school level, which in turn stems from the clash at the system level.

The professional opportunities conflict (A) on the education system level (A1) is expressed in two ways:

(a) **Limited opportunities for career advancement**

Analysis of the interviews attested to the importance the teachers attribute to their personal development and professional advancement while, in fact, the only possibilities for promotion and advancement lie primarily in the administrative track. Thus, STEM teachers are in conflict with the education system, which in their perception fails to provide them with opportunities for advancing their *teaching* careers.

The STEM teachers interviewed in this study were more interested in the development of the professional aspects of their careers and less so in terms of the administrative aspects. In particular, the STEM teachers who participated in the study expressed the desire to fill research and development (R&D) positions, in which they would engage in the development of curricula, serve as the teachers of teachers, study toward their PhDs, and fulfill leadership roles within the education system.

For instance, D., a math teacher, said: *Teachers do not have many opportunities. Taking myself as an example… Had I not found an opportunity in the computerization of the network (a national education network the school belongs to), I would not be in teaching any more… It did not take long before I found myself in a management course, where I discovered that was not what I wanted… **Part of the teacher's position should be R&D… in other words, teachers should not only teach but rather they should engage in some sort of research, and also develop study programs that will serve as the basis (for curriculum materials)** [emphases and punctuation by the authors].*

The teachers' words, as expressed in the interviews, raised the question of whether or not STEM teachers only rarely fill administrative positions. The answer to this was clarified by means of an online questionnaire distributed in the schools participating in the study.[5] It was found that only a small percentage of teachers teach the five science subjects examined in the study fill educational and administrative positions. On average, 16.8% of all teachers in the school are STEM teachers and only 8.32% of all administrative positions-holders in the school are STEM teachers (Appendix 3 presents the data in detail).

It is important to mention that even if teachers wish to advance into administrative roles, the number of such roles open to teachers is limited (Barak 2011). This stems, among other things, from the flat structure of the educational organization (Oplatka 2007).

One alternative for minimizing conflict in organizations of this kind is to develop a dual development ladder, mentioned above, that enables two tracks of development: a professional track and an administrative track. In Israel, an attempt has been made in recent years to implement a similar policy by means of the "New Horizon"[6] reform, which presents teachers with successive stages of development so that teachers at advanced stages are appointed to defined leadership positions, which bear influence on the entire system. In the "Courage for Change" reform, teachers are required to undergo pre-training prior to their appointment to management positions (such as class coordinator or subject coordinator). However, such requirement does exist for professional leading roles. Thus, the needs of STEM teachers for a professional career horizon are not met.

In general, the constitutive documents of the "New Horizon" and "Courage for Change" reforms, which both aimed at enhancing the quality of teaching and improving student achievements, indicate that the education system views the issue of teachers' professional development mainly from a standpoint of enriching teachers' professional knowledge, with less emphasis on opening up advancement options in the professional track. In other words, these reforms emphasize disciplinary expertise more than the development of teaching careers and changing the substance of the profession.[7]

(b) The need for a cooperative professional community

Though the Minister of Education offers many professional development programs as well as national teacher centers, the STEM teachers expressed the need for continual professional courses as well as a community of teachers that is external to the teacher's own school. The need to belong to a professional community is one

---

[5]The questionnaire was sent to school. One of the managerial role holders filled it.

[6]New Horizon (Ofek Hadash) is an educational and professional reform in elementary and junior high school education. Its implementation began in 2008.

[7]From the Ministry of Education Web site: A letter to principals "Policy and Guidelines for Making a Plan for the Professional Development of Teachers" sent by Motti Rosner, Director of Department A, Professional Development of Teachers, June 2012. Retrieved from http://cms.education.gov.il/educationcms/units/pituachmiktzoie/oz/hearchut5773.htm (In Hebrew).

characteristic of a professional's professionalism (Hall 1968). Since this need remains unfulfilled, a conflict arises between the professionals and the organization, which is followed by a decline in the professional's organizational commitment and job satisfaction (Shafer et al. 2002).

The professional opportunities conflict (A) on the school level (A2) is expressed by **the need for professional development and for a cooperative professional community**

The conflict described above (A1) on the education system level is manifested on the school level as well, where teachers are in need of professional interaction with a supportive professional team for the purpose of exchanging material, writing examinations, sharing knowledge, and more.

According to Barak (2011), organization-oriented policy assumes that in order to enhance the quality of teaching and learning, teachers need to support each other, develop, learn together, and collaborate. In interviews conducted within the framework of this study, the conflict was manifested in weaknesses on the school level that the teachers mentioned. The teachers reported professional isolation manifested by an absence of teamwork, in the most practical sense of the word, which could help improve teaching methods, through the sharing of learning materials.

The professional opportunities conflict (A) on the student level (A3) is exhibited by **occupational burnout and pedagogical difficulties**

The conflict on this level is an outcome of the conflict as it is manifested on the two higher levels (system and school). This conflict relates to teaching in the classroom and to the teacher-student encounter.

The scarcity of opportunities for advancement in the organization, teaching careers whose end is almost identical to their commencement, the teaching of large classes, and the routineness of the work all lead to teachers' professional burnout. In general, burnout is expressed by fatigue, solitude, impatience, and dissatisfaction with oneself and with colleagues (Friedman and Gavish 2003). The professional theory indicates a strong correlation between lack of opportunities for professional advancement and dissatisfaction and burnout at work (Herzberg 1966; Sergiovanni and Carver, 1980 in Oplatka 2007). In particular, analysis of the interviews indicated that the teachers perceive this occupational burnout as a personal shortcoming, while expressing concern that the students are those who suffer from it. Here is one illustrative quote:

D., a math teacher, said: *The main threat is that I'll get fed up* … *I'm getting a little tired of explaining the same thing… a little impatient from doing the same things, over and over again… it's hard, going over and over the same thing several times. This is a weakness… and those who suffer from it are mainly the students.*

The lack of both a cooperative professional community and relevant continuing education courses, which contribute to bolstering and updating the teachers' knowledge components and to improving teaching methods, jeopardizes the teachers' level of professionalism and leads to weaknesses on the pedagogical level. The teachers also mentioned pedagogical difficulties, such as the need for a variety of teaching methods and the need to expand their disciplinary knowledge. Indeed, they noted these difficulties as personal shortcomings, but also as such that have a detrimental effect on teaching quality, which mainly harms the students.

1.4.2 The Profession Perception Conflict (B)

This conflict embodies the clash between the STEM teacher's perception of how scientific subjects should be taught and that of the organization. Such a confrontation is created when the organization operates in a way that contradicts the characteristics of the teacher's professionalism: their social commitment, their devotion to the profession, their belief in self-supervision, and their demand for professional autonomy (Hall 1968).

The perception of the teachers who participated in the study regarding the teaching of science and its importance relates to aspects such as the desirable number of teaching hours allocated for the subject, the content to be taught and the appropriate teaching methods, and the level of student commitment to the subject. Teachers are compelled to obey the organization's requirements, and when these contradict their views, the said conflict arises, as described in the following.

The profession perception conflict (B) on the education system level (B1) stems mainly from **the structure and content of STEM education in junior high schools**

Analysis of the data attests to the high school teachers' attitude toward STEM education in junior high schools; more specifically, high school STEM teachers criticized the way in which STEM education is perceived and implemented in the junior high schools. The teachers pointed out two weaknesses in this context. The first is on the curricular level: They believe that the different science subjects (physics, chemistry, and biology) should be taught individually as distinct subjects, and not as a single, unified "sciences" subject that includes physics, chemistry, and biology. The second weakness, which perhaps derives from the first, relates to pedagogical mistakes made in teaching the subject in junior high school. These mistakes stem from the fact that the teachers are experts in only one area of knowledge (i.e., physics, chemistry, or biology), whereas they are required to teach the entire range of science subjects.

The teachers are therefore in conflict. On the one hand are the professional characteristics such as social commitment and devotion to the profession (Hall, 1968) and on the other hand is the dictate of the education system to teach the three science subjects as one subject. According to the teachers, the way in which the teaching of STEM in junior high schools is structured constitutes a threat to the state of STEM education in general. They expressed their concern that students reach high school with only meager and inaccurate knowledge of the three science subjects and, as a result, may not choose to focus on them in their matriculation examinations.

The profession perception conflict (B) on the school level (B2) is manifested in the **loss of teaching hours due to school activities**

Once again teachers find themselves at odds with the school, as many teaching hours are canceled due to unscheduled school activities. On the one hand, teachers are committed to professionalism (Hall 1968), while on the other hand, they must comply with the dictates of the school administration. The loss of teaching hours jeopardizes the achievement of goals set and necessarily compromises the teachers' professionalism. The teachers presented this problem as a weakness on the school level.

Thus, according to A., a physics teacher: *Many hours are lost... when you check objectively... how many hours the curriculum is supposed to give as opposed to how many you actually have, so it's a problem.* [Though the school end at the end of June, in practice] *the school year ends at Passover* [two months earlier]. *From the tenth grade on, school ends after Passover.*

The profession perception conflict (B) on the student level (B3) is manifested in two ways:

(a) **Student obligations in school**

On the student level, too, teachers find themselves at odds with the system in all that pertains to the perception of the profession. On the one hand, the teachers' priority is to focus on the pedagogical aspect, i.e., teaching the science subject, that stems from their professional commitment (Hall 1968), while on the other hand, the system encourages (and at times even initiates and requires) a whole range of other student activities such as community service. This overloading of the student's schedule with additional activities and commitments causes them to miss class hours, and they are unable to fulfill the teachers' curricular requirements. The teachers note this in the context of the weaknesses of the school, which does not prioritize the STEM education over other obligations.

(b) **Dealing with students' discipline problems and personal matters**

Dealing with student matters occupies a large part of the teacher's daily routine. This contradicts the STEM teachers' perception of the profession. The need to enforce discipline on unmotivated students only increases the burden and stress on the teachers, thus making it harder for them to fulfill their professional duties (Oplatka 2007 p. 185).

1.4.3 The STEM Education Discourse Conflict (C)

In Israel, not many high school students choose to study science (physics, chemistry, and biology) and mathematics at the highest level (a five-point level). Ministry of Education data for 2014 show that the percentages of students sitting for the matriculation examinations in the sciences (for highest level) from the total number sitting for matriculation examinations in all subjects were 19% in biology, 9% in physics, 8% in chemistry, and 10% in the highest level of mathematics.[8] These data reflect on the number of young people advancing to higher education in these areas. Data from institutions for higher education show the following distribution of undergraduate students in the sciences (the year 2011): mathematics, statistics, and computer sciences 5%; physics 1.3%; biology 2.6%.[9]

As professionals, STEM teachers exhibit a high level of professional devotion to the teaching and possess a high level of social commitment (Hall 1968); in their opinion, the way in which the organization operates contradicts their social commitment as people who are encouraging students to enter the world of science learning. Thus, according to the teachers' understanding, when the education system does not project the importance of STEM education to school principals and through them to the students, the number of students who choose to study science declines, and their conflict with the organization exacerbates.

Specifically, **the STEM education discourse conflict (C) on the education system level (C1)** is expressed by the importance attributed to the percentage of matriculation eligibility versus the composition of the matriculation certificate. Till recently, the Ministry of Education measures the success of schools according to the matriculation examination score data (entitled/not entitled and grades relative to the national average) rather than by the level of study and the cluster of subjects the pupils study.[10] The result is that the perception of success does not emphasize the significance of the STEM subjects. It should be stressed that students perceive the science subjects (physics, chemistry, and biology), mathematics, and computer science (all on a five-point level) as more difficult and so they are more reluctant to choosing to study them. This policy regarding the measurement of school success leads principals, teachers, homeroom teachers, and school counselors sometimes to direct students to give up studying STEM subjects, thus improving both the students' matriculation averages and the matriculation eligibility percentage for the school. According to the STEM teachers, if school success was judged using other

---

[8]Data were informed to the researcher directly from the subject coordinator in the Ministry of Education.

[9]The Council for Higher Education (CHE) Report: Collection of Data, Article 14, Table 3. Retrieved from http://che.org.il/wp-content/uploads/2012/05/%D7%9C%D7%A7%D7%98-%D7%A0%D7%AA%D7%95%D7%A0%D7%99%D7%9D-%D7%AA%D7%A9%D7%A2%D7%91.pdf (In Hebrew).

[10]This situation has been changed in 2016, when high schools were started being measured also by additional factors which refer to the quality of the matriculation examination. These measures are captured in the school educational picture.

criteria, it would be possible to increase the number of students who choose the STEM subjects.

The STEM education discourse conflict (C) on the school level (C2) is exhibited, according to the STEM teachers, by the fact that **the school's message regarding the importance of STEM education is incoherent**. When school management fails to encourage discourse that stresses the importance of STEM education, it results in few students choosing to study scientific subjects at an advanced level. This too puts the STEM teacher at odds with the school. The teachers mentioned this as a weakness on the school level, whereby, as mentioned above, the system does not measure a school's success according to the quality of its STEM education, thus denying it the incentive to develop a positive discourse on the subject.

A., a physics teacher, presents his position on the matter:

> I think that the school does not do enough to encourage the study of physics, not only physics, hard sciences... in order for students to study physics, there needs to be an atmosphere within the school that physics is important... it needs to be broadcasted by the system. The system is the administration, the coordinators, the homeroom teachers. There is a total misunderstanding on the part of the entire administrative team.

The STEM education discourse conflict (C) on the student level (C3) is expressed in **the need to "market" the subject to students**. As an outcome of the expression of the STEM education discourse conflict at the education system and school levels, and in light of the percentage of students choosing to study STEM subjects at an advanced level, STEM teachers are required to "recruit" their students and "market" their specialty subject, in order to preclude the shutting down of the school's science track. On the one hand, STEM teachers are committed to their profession and to society (Hall 1968); on the other hand, they are subjected to pressure by the organization (Shafer et al. 2002) and are required to support organizational goals and be involved in marketing the subject matter they teach.

L., a computer science teacher, explains the expression of this situation in her school:

> The number of students is limited...Physics has a certain prestige about it which lures them, and the rest of the subjects have to fight for the students, and then it's a question of who does better marketing and I personally have difficulty playing that role.[11]

1.4.4   The Academic Freedom Conflict (D)

The professional teacher's need for academic freedom is in keeping with a central characteristic of professionalism, i.e., the demand for professional autonomy (Hall

---

[11]These words, though not necessarily reflect the real situation, expressed this computer science teacher' perception with respect to physics classes. Data shown in the beginning of the section emphasize that on the country level many pupils also do not choose to study physics on the highest level.

1968). The more rigid the system and the more rigid its orders and dictates, the less academic freedom is experienced by the professionals who work in the organization.

Specifically, in the context of the research presented in this paper, **the academic freedom conflict (D) on the education system level (D1)** is manifested by **the lack of autonomy in curricular issues**. In the weakness dimension, the teachers noted that academic freedom was denied them in many areas: choosing study contents and their level, determining the number of annual teaching hours allocated to the subject, and the level and structure of the matriculation examination. Furthermore, teachers presented the need to transition to online units in matriculation examinations, the lack of which makes it difficult to incorporate technology-assisted teaching, as an opportunity on the curriculum level. Finally, teachers also raised the need to synchronize teaching of study materials among the various subjects.

1.4.5   The Teacher's Status Conflict (E)

The professional advancement of teachers, their salary, social benefits and pension, as well as working environment, are all determined by the education system (or local authority) and are all closely connected to the attractiveness and image of the teaching profession.

The 2011 OECD report titled "Building a High-Quality Teaching Profession" (OECD 2011) reports on countries that have succeeded in turning teaching into an attractive profession, by strengthening the teacher's status. These countries have achieved this by taking two main measures: (a) creating a promising employment prospect for interesting teaching careers and (b) providing high-quality teacher training. In other words, the report mentions education systems that recruit high-quality teachers by creating an environment in which they work as professionals. In countries where this policy is implemented, it has proven to have had considerable influence on the allure of the teaching profession. According to this OECD report, attractive conditions enhance morale, reduce retirement, and enlarge the reservoir of teachers.

In the study presented in this paper, this conflict is manifested both on the education system level and on the school level.

Teacher's status conflict (E) on the education system level (E1) exists **between the teachers' self-perception as STEM teachers and professionals—** people who have been trained for their jobs in academic institutions, hold academic degrees in the subjects they teach, and wish to continue their professional development—**and the way in which the system perceives them (and to a large extent society as a whole)**. According to the STEM teachers who participated in the study, teachers' employment conditions (their salary and work environment) reflect the way they are perceived by the system. The teachers mentioned the salary and pension levels as a personal threat, as components of their job that jeopardize their chances of remaining in the teaching profession.

Teacher's status conflict (E) on the school level (E2) is manifested by the STEM teachers' perception of **their working conditions** as a weakness on the school level: poor laboratory conditions, lack of equipment, and lack of sitting alcoves and rooms for the professional teachers.

R., a biology teacher, comments:

> *It is too crowded here. Imagine if I had to conduct an experiment in here; then I have a problem. If I have a class of 30, I can't do an experiment with them because they're so many, and it isn't safe… the physical means - they must be improved.*

1.5   Summary and Discussion

The aim of the research described in this paper was to characterize STEM teachers' perceptions of STEM education in the Israeli school system. Analysis of the data indicated that the STEM teachers' perception may be presented through five conflicts that exist between the STEM teachers and the educational organization to which they belong: **A. the professional opportunities conflict; B. the profession's perception conflict; C. the discourse on STEM education conflict; D. the academic freedom conflict; and E. the teacher's status conflict**. These conflicts are manifested on three different levels (the system, the school, and the students), whereby conflicts on the school level and the student level stem, at times, from the conflict on the system level. Furthermore, it can be seen that all of the conflicts, even when not manifested on all levels, are always expressed on the education system level (Table 1.1). It may therefore be concluded that STEM teachers are in conflict mainly with the organization (in its broader sense) in which they are employed, i.e., the education system, more than they are with the school or the students. Thus, the study should be expanded, and the question of whether first to deal with managing the conflicts on the system level should be addressed.

Conflicts A–D were ascribed to the conflict known in the sociological–organizational literature as the bureaucratic–professional conflict, which arises when the professional values of teachers-professionals clash with the values of the bureaucracy in general. The bureaucratic–professional conflict leads to a situation in which the professionals encounter difficulty in realizing their goals (Hall 1967), which in turn results in the organization's difficulty to achieve its objectives. Here lies the significance of this study whose findings may offer a new viewpoint regarding the challenges STEM education in Israel faces.

Moreover, the conflict over the need for professional development and a cooperative professional community (A2) as well as the burnout and pedagogical difficulties conflict (A3) indicates the importance teachers attribute to broadening their professional knowledge and expanding the variety of teaching methods. As a result, future studies should propose renewed planning of continued education courses for STEM teachers, as well as the establishment of a professional community of STEM teachers.

The teacher's status conflict (E) does not necessarily stem from the professional values of the teachers as professionals but rather from their expectations as employees in an organization. In the present study, the teachers expect the organization to reinforce the status of the teaching profession in society. This conflict, together with the limited opportunities for career advancement conflict (A1), can explain the difficulties encountered in recruiting new STEM teachers as well as the teachers' difficulties persisting in their jobs. According to the OECD report (OECD 2011), providing a favorable prospect for professional development, as well as delegating responsibility to the teachers as professionals, is important parameters that are likely to be manifested in an increase in the allure of the teaching profession.

Therein, our findings reinforce the feeling prevalent in the public in general and in the education system in particular, regarding the need for a proactive approach in positioning and promoting a change in STEM education in Israel.

Appendix 1 The Professional and Academic Background of the STEM Teachers

Name of teacher (initial) and subject	School nickname	Teaching experience (years)	Academic background	Additional roles in and out of school
1. K.—Chemistry	A	20	B.Sc. & M.Sc. Chemistry + Teacher Certificate	Chemistry coordinator, science cluster coordinator
2. S.—Chemistry	B	18	B.Sc. Chemistry + Teacher Certificate, M.Sc. Medical Sciences (TAU[a])	Homeroom teacher; chemistry coordinator
3. R.—Biology	A	32	B.Sc. Life Sciences (HUJI[b]) + Teacher Certificate	Homeroom teacher
4. P.—Biology	B	31	B.Sc. Biology + Teacher Certificate (TAU)	School feedback supervisor; final project and courses coordinator; former high school principal
5. D.—Mathematics	C	12	B.Sc. & M.Sc. Computers & Math Education, Ph.D. in Science Education	Manager of IT system in national network to which her school belongs

(continued)

(continued)

Name of teacher (initial) and subject	School nickname	Teaching experience (years)	Academic background	Additional roles in and out of school
6. I.—Mathematics	C	2 (after career change for academics)	B.Sc. Math & Computer Sciences currently studying for M.Sc. in Math & Computer Science Education	Math coordinator; instructor in online teachers courses on use of technology in math education; background: 10 years in high tech; participated in academics career change program in 2010
7. T.—Mathematics	E	22	B.Sc. Mech. Eng. (TAU) currently studying for M.Sc. in Science Education	No additional roles in the school; background: worked in industry; teacher since 1990; 19 years teaching at the school
8. M.—Mathematics and computer sciences	E	21	B.Ed. (teachers college)	Mathematics coordinator; homeroom; teaches religious subjects
9. L.—Computer sciences	A	17	B.A. Social Sciences	Computer science coordinator; coordinator of software engineering track; background: participated in academic career change to computer science teacher in 2009
10. A.—Physics	F	15	B.Sc. & M.Sc. Electrical Engineering (TAU)	Physics coordinator; serves as pedagogical coordinator in private NPO outside the school; provides counseling for university students; instructor in preparatory courses for the Physics Olympics

(continued)

(continued)

Name of teacher (initial) and subject	School nickname	Teaching experience (years)	Academic background	Additional roles in and out of school
11. B.—Physics	D	23	B.Sc. & M.Sc. Physics; B.Sc. Metallurgy + Teacher Certificate	Physics coordinator; works at another school as well
12. M.—Physics	A	28 (18 years in Israel)	B.Sc. & M.Sc. Physics + Teacher Certificate	Physics coordinator; works at another school as well

[a]Tel-Aviv University
[b]Hebrew University of Jerusalem, Israel

Appendix 2 Examples of Career Ladders in Education Systems Around the World

From the OECD report (OECD 2011): "Building a High-Quality Teaching Profession"

- Australia: 2–4 career stages are offered with a pay raise at each stage promotion from novice to experienced teacher, from experienced leading teacher to vice-principal, and finally, from vice-principal to principal or regional/district official. Each stage requires a higher level of knowledge, effective teaching, assuming of responsibilities within the school framework, assisting colleagues, and so on.
- England and Wales: Outstanding teachers whose qualifications have been evaluated are offered leadership roles and professional development within the school and in other schools. These roles may constitute up to 20% of the teaching position.
- Ireland: Four promotion categories: principal, vice-principal, assistant principal, and teacher with special duties. Each category has its own areas of administrative responsibility, level of salary, and time allocation.
- Canada and Quebec: Experienced teachers are promoted to student mentoring roles with either an increase in salary or a decrease in the scope of their position as classroom teacher upon assuming the additional role.

Appendix 3 Percentage of STEM Teachers in Educational and Administrative Positions

Post-primary school	Total teachers in the school		Homeroom teachers		Class (year) coordinators		Other administrative positions[c]	
	Proportion of STEM teachers[a] out of total teachers (%)	Proportion of teachers of all other subjects[b] out of total teachers (%)	Proportion of STEM teachers[a] out of total homeroom teachers (%)	Proportion of other teachers out of total homeroom teachers (%)	Proportion of STEM teachers[a] out of total coordinators (%)	Proportion of other teachers out of total coordinators (%)	Proportion of STEM teachers[a] out of administrative positions (%)	Proportion of other teachers[b] out of total admin. positions (%)
A	28	72	19	81	33.3	66.6	12.5	87.5
B	16	84	23	77	0	100	16.6	83.5
C	10	90	33.3	66.7	0	100	0	100
D	10	90	13.3	86.7	0	100	12.5	87.5
E	20	80	None[d]	100	None	100	None	100
F					0	100	14.28	85.72
Average	16.8	83.2	22%	88	6.6%	93.4	8.32%	91.68

[a]STEM teachers teach mathematics, physics, biology, chemistry, and computer sciences in grades 10–12 and prepare them for three-point (lowest) to five-point (highest) level matriculation

[b]Teachers of all other subjects: teachers who do not teach any of the aforementioned STEM subjects

[c]Administrative positions: school principal, vice-principal, high school grade administration, pedagogical coordinator, or any other administrative position

[d]At this school, only teachers of religious subjects fill the positions of homeroom teachers, class coordinators, and administration

References

Afzalur Rahim, M. (2002). Toward a theory of managing organizational conflict. *International journal of conflict management, 13*(3), 206–235.

Barak, M. (2011). Improving the quality of teachers by means of organization-oriented policy: The whole (school) is greater than the sum of its parts (teachers). *Hed Hahinuch, 86*(1), 29–30. Retrieved from http://portal.macam.ac.il/ArticlePage.aspx?id=4471 (in Hebrew).

Cesare, S. J., & Thornton, C. (1993). Human resource management and the specialist/generalist issue. *Journal of Managerial Psychology, 8*(3), 31–40.

Friedman, I., & Gavish, I. (2003). *Teacher burnout: Shattering the dream of success*. Jerusalem: The Henrietta Szold Institute. (in Hebrew).

Hall, R. H. (1967). Some organizational considerations in the professional-organizational relationship. *Administrative Science Quarterly*, 461–478.

Hall, R. H. (1968). Professionalization and bureaucratization. *American Sociological Review*, 92–104.

Hall, D. T. (2002). *Careers in and out of organizations* (Vol. 107). Beverly Hills: Sage.

Hativa, N. (2008). Is teaching a profession?. *Hed Hahinuch*. Retrieved from http://www.itu.org.il/?CategoryID=1476&ArticleID=12506 (in Hebrew).

Herzberg, F. I. (1966). *Work and the nature of man*. Crowell Co: Thomas Y.

Initiative for Applied Education Research. (2012). *What do those engaged in the teaching of mathematics in post-primary school need to know?* The Israel Academy of Science and The Humanities. Retrieved from http://education.academy.ac.il/Admin/Data/Publications/Math-Final-Report.pdf (in Hebrew).

Noe, R. A. (2002). *Employee training and development*. Boston, MA: McGraw-Hill/Irwin.

Oplatka, I. (2007). *Fundamentals of education administration: Leadership and management in the educational organization*. Pardess Publishing (in Hebrew).

OECD. (2011). *Building a high quality teaching profession: Lessons from around the world*. OECD Publishing. Retrieved from http://www2.ed.gov/about/inits/ed/internationaled/background.pdf.

OECD. (2015). *Education at a Glance 2015 OECD Indicators*. Retrieved from http://meyda.education.gov.il/files/MinhalCalcala/EAG2015.pdf.

Patton, M. (1990). *Qualitative evaluation and research methods*. Beverly Hills, CA: Sage.

Shafer, W. E., Park, L. J., & Liao, W. M. (2002). Professionalism, organizational-professional conflict and work outcomes: A study of certified management accountants. *Accounting, Auditing & Accountability Journal, 15*(1), 46–68.

PCAST. (2010). *Prepare and Inspire: K-12 Education in Science, Technology, Engineering, and Math (STEM) for America's Future*. Retrieved June 21, 2017, from http://stelar.edc.org/publications/prepare-and-inspire-k-12-education-science-technology-engineering-and-math-stem.

Wilensky, H. L. (1964). The professionalization of everyone? *American Journal of Sociology*, 137–158.

World Economic Forum. (2012). *The Global Competitiveness Report 2012–2013: Country profile highlights*. Retrieved June 2, 2017, from http://www3.weforum.org/docs/CSI/2012-13/GCR_CountryHighlights_2012-13.pdf.

Chapter 2
SWOT Analysis of STEM Education in Academia: The Disciplinary versus Cross Disciplinary Conflict

Yehudit Judy Dori, Tali Tal & Einat Heyd-Metzuyanim

Abstract In this chapter, we show how to turn a SWOT analysis process of an academic faculty into a strategic plan in a way that fosters the strengths, eliminates the weaknesses, exhausts opportunities, and deals with outside strengths. The case described in this chapter focuses on the Technion's Faculty of Education in Science and Technology. The SWOT characteristics, though specific to this particular faculty, are also a result of the tensions that exist almost in any universally between teaching and investigating *disciplinary* content knowledge on the one hand and the application of *general* pedagogical principles, approaches, and methods on the other hand. On top of these tensions, there are also constraints of funding, human resources, and workload of both faculty and students, which are common to all the academic disciplines. As such, the analysis can serve as a useful managerial tool for coping with the challenges that a Science, Technology, Engineering, and Mathematics (STEM) faculty faces while finding creative avenues for growth and improvement of their teaching and research.

Keywords Academia · Disciplinary vs. interdisciplinary · STEM education · SWOT analysis

2.1 Introduction

This chapter focuses on how to find effective ways for managing an academic STEM education faculty for delivering STEM education at the highest possible level in the face of a lingering conflict between disciplinary focus and interdisciplinary orientation.

There are several models for promoting STEM education. One is STEM discipline affiliation, i.e., science educators are research faculty who are affiliated with the respective STEM departments, that is, a mathematics educator in the math department, a biology educator teaches and conducts research as a faculty member in the biology department, and so forth. Examples of universities that apply STEM affiliation include North Carolina State University (e.g., Robert Beichner—a physics educator at the Department of Physics); University of Massachusetts Boston (e.g., Hannah Sevian—a chemical educator at the Department of Chemistry), or University of Arizona (Vicente Talanquer—a chemical educator at the Department of Chemistry and Biochemistry). Another category under this model is a "second career in STEM education"—scientists who chose to focus on science education research and practice in addition to "pure science" (e.g., Carl Edwin Wieman a physicist and educationist at Stanford University, who was awarded the Nobel Prize in Physics; Joe Redish at the Department of Physics at Maryland University; John Belcher at the Department of Physics at MIT).

A second model is STEM education affiliation in which a group of STEM educators operates as an independent unit or as part of a school or college of education. Technion, Weizmann Institute, Michigan State University, and University of Georgia are examples of the STEM education affiliation model.

Each of the models—the STEM affiliation model and the STEM education affiliation model—has advantages and challenges presented in Table 2.1.

In this chapter, we present and analyze a case study involving the Faculty of Education in Science and Technology at the Technion. This STEM education academic unit is unique, since it is equivalent to a school in the US university system, and it is the one of a few that covers every STEM discipline.

Focusing on this academic unit, we carry out a detailed SWOT analysis to identify advantages and challenges with respect to delivering STEM education to almost all age levels at the highest quality possible.

Table 2.1 The STEM affiliation model and the STEM education affiliation model—advantages and challenges

Educational model	Characteristics	Advantages	Challenges
STEM affiliation	Focus mostly on higher education	Possible collaboration with faculty teaching the courses; immediate access to research population and settings	Often one science educator is alone in a discipline; little interaction with other STEM educators at the same university
STEM education affiliation	Focus on investigating various age groups	A group of science educators who can collaborate on big projects	Distance from some research settings, e.g., schools
		Focus on broad areas of research: learning, teaching, informal education, out-of-school education, science communication, educational technology	

2.2 Theoretical Background

As noted in the introduction, science education scholars and researchers affiliation follows one of two main models: STEM affiliation, in which the science educator is associated with a mathematics, science, or engineering department, or the STEM education affiliation, in which the science educator is associated with a school, college, or department of education. The Policy Forum of Science Magazine (Handelsman et al. 2004) has emphasized scientific teaching, which involves active learning strategies to engage students in the process of science and teaching methods that have been systematically tested and shown to reach diverse student populations. The authors explicitly noted: "*Universities need to provide venues for experienced instructors to share best practices and effective teaching strategies. This will be facilitated, in part, by forming educational research groups within science departments. These groups might be nucleated by hiring tenure-track faculty who specialize in education…*" (p. 522). Almost a decade later, the Discipline-Based Education Research (DBER) Report (Singer et al. 2012) acknowledged the field of education research in higher education and recommended that "*science and engineering departments… clarify expectations for DBER faculty positions.*" (p. 4). However, much research on teaching science and engineering is done by science educators affiliated with dedicated academic units that focus on science education which may be independent, like the Faculty of Education in Science and Technology at the Technion, or located within a school or a college of education, as is the case with the School of Education of University of Colorado Boulder or the College of Education at the University of Missouri-St Louis and many more such departments.

STEM education studies are collaborative efforts of STEM educators and scientists. Such initiatives have been described in (a) Two Special Issue of Journal of Science Education and Technology—JOST focusing on Educational reform at MIT (Dori 2007, 2008) and (b) the Special Issue of the Journal of Research in Science Teaching—JRST on Discipline-Centered Post-Secondary Science Education Research, co-edited by a scientist and a science educator (Coppola and Krajcik 2014). Indeed, most papers in the JOST special issues and the one of the JRST Special Issue were written collaboratively by scientists and science educators or engineering educators and engineers. Such collaborations occur regardless of affiliation and position (e.g., Barak et al. 2007; Dori and Belcher 2005; Dori et al. 2007; Lipson et al. 2007; Marbach-Ad et al. 2010; Mintz et al. 2013; Mitchell et al. 2011; Tsaushu et al. 2012), but in general, science educators are more focused on K-12 rather than on post-secondary teaching and learning.

Examining the two models of STEM educators' affiliation, the question that arises is whether the structure matters. Biglan (1973) found that social connectedness was significantly higher among scientists ("hard areas") than among social scientists (and other "soft areas"). In contrast, researchers in "soft areas" are more committed to teaching. Such distinctions imply that scholars in different fields have unique and different cultures. Becher and Trowler (2001) referred to mapping of "Tribal Territories" and use the metaphor of landscape to emphasize that these territories are not only natural, but also socially constructed. They argued that in light of differences between academic territories and landscapes, attention should be paid to differences between disciplinary and even to subdisciplinary groups. Hativa and Marincovich (1995) acknowledged the pioneering work of Biglan and his analysis of $2 \times 2 \times 2$ matrix of hard versus soft areas, applied versus pure areas, and life versus non-life area. They argued that in addition to the recognition of differences between disciplines, there is growing recognition of the differences in learning within the disciplines. Despite the increasing awareness of the power of disciplines, there is relatively little research on the boundaries between them and their characteristics in terms of quality of teaching and learning in higher education.

Given this state of affairs, the question raised is what possible ways of operating and managing faculty and experts would provide high-quality STEM education?

The focus of the analysis provided hereinafter is on the Faculty of Education in Science and Technology at the Technion, Israel Institute of Technology. This Faculty is an independent science and technology education faculty (school in US terms), which is situated within a science and engineering research university. Using Biglan's terms, the question being asked is: What is the status of an academic unit that does "soft" research on teaching and learning in "hard" disciplines?

Using Becher and Trowler metaphors, we look at landscapes while using the ecological metaphor of relationships.

2.3 Setting

The Faculty of Education in Science and Technology at the Technion, Israel Institute of Technology, is one of 18 academic units (mostly Faculty). With 12 faculty members including the dean, the mission of the Faculty of Education in Science and Technology (hereinafter "the Faculty") is to pursue excellence and maintain international leadership in science, technology, engineering, and mathematics (STEM) education. The Faculty conducts this mission through our highly innovative research on learning, instruction, and assessment in both formal and informal educational settings and across the learner's life span. The Faculty is the largest of its type in Israel and one of the largest worldwide. It has 12 full-time faculty members, six of whom are tenured and six in tenure-track positions. The faculty members are distinguished, established scholars and leaders in their respective STEM education communities. Their achievements include extensive lists of publications and participation in international and national professional fora, including invitations to give keynote lectures in conferences and seminars.

2.3.1 Structure, Organization, and Research of the Faculty

The Faculty is structured to support our mission by achieving excellence in research, educating prospective teachers and training in-service teachers, mentoring of future generations of researchers, and sustaining communities of practice in STEM teaching and learning. Dean (first author of this chapter) heads the Faculty. In addition to the 12 faculty members, there are about 30 adjunct teaching specialists and nine administrative and technical staff. The Faculty's academic programs cover learning and teaching of all the STEM topics within the context of six thematic areas: teaching and teacher education, curriculum, learning environments, lifelong learning (LLL), science communication, and educational neurosciences. Educational research methods are also a distinct teaching theme. Faculty members are responsible for the academic programs and core teaching activities in all the themes.

Basic and applied research within the Faculty covers both the traditional fields of STEM education, such as learning, teaching, and assessment, and new areas, such as learning sciences, neuroimaging and neuro-education, science communication, multicultural education, ethics education, and entrepreneurship. Our research spans these STEM topics, as represented in the Faculty's research portfolio. Faculty members and their graduate students publish in leading journals and collaborate extensively both nationally and internationally. In addition to formal and informal science learning environments, faculty members conduct applied research that takes place in and informs K-12 education, higher education, the public and private sectors, and third sector organizations.

2.3.2 Academic Programs at the Faculty as a Way to Achieve Excellence in STEM Education

Recalling that the goal of this chapter is to propose a way to organize an academic unit for achieving excellent STEM education, in this section, we describe the six academic programs at the Faculty and how their organization serves the purpose we seek to achieve.

The Faculty offers several programs for a large body of students: (a) a four-year B.Sc.Ed. program, (b) Views I program in STEM Education for Technion STEM graduates, which is a postgraduate program,[1] (c) M.Sc. with and without research thesis, (d) Views II program—M.Sc. with teaching certificate for excellent Technion STEM graduates, (e) a Ph.D. program, and (f) a teaching certificate for non-Technion graduates. Orthogonal to this cross-disciplinary set of programs,

---

[1]The Council for Higher Education (CHE) considers graduates of this program as completing an additional B.Sc. degree.

studies are organized around eight disciplinary education tracks: biology, chemistry, computer science, electrical engineering, environmental sciences, mathematics, mechanical engineering, and physics. Additionally, there are six cross-disciplinary themes, including educational neuroscience, educational technology, science communication, science education in informal environments, project assessment, higher education, and career paths.

Accounting for the 14 tracks (comprising the eight disciplinary and six cross-disciplinary tracks) on one hand and the six study programs, there are 84 cells in a matrix, where each cell is a theoretical combination of a study program in some track. However, in practice, students from different tracks and level of maturity in the research and non-research tracks are studying together. One example is integrated courses in the B.Sc.Ed. and M.Sc. programs in Mathematics Education. Another is the M.Sc. and the Ph.D. program in Electrical Engineering Education.

This matrix organization, while difficult to manage, provides at least a partial solution to the inherent tension existing between general educational principles—pedagogical knowledge—and domain-specific or disciplinary knowledge—pedagogical content knowledge (PCK, Shulman 1986), i.e., pedagogical knowledge that is specific to each STEM domain and needs to be studied and applied separately for almost each of our eight disciplinary educational tracks. Moreover, the six programs span a wide spectrum of expertise levels, from the basic one that B.Sc.Ed. students get all the way to the Ph.D. programs. This variability requires adjusting the level of teaching and learning in order for them to be in par with the diverse capabilities and backgrounds of the students enrolled in these programs.

How is this feasible? How can a faculty body of 12 members control such a bewildering variety of programs, disciplines and cross-disciplinary domains? This is indeed a huge challenge, which is discussed as part of our SWOT analysis, described in the next section.

2.3.3 The Method: SWOT Workshop

SWOT analysis highlights the strengths and weaknesses of an organization, as well as the opportunities and threats it faces (see more details in Chap. 1 of this Brief as well as Afzalur Rahim (2002). This type of analysis facilitates the development of strategies that amplify the strengths, maximize the advantages of opportunities, and defend the organization from weaknesses and threats.

Strengths are characteristics of an organization that give it advantages over others. Weaknesses are characteristics that place the organization at a disadvantage. Opportunities are phenomena, generally external to the organization that can be used to its advantage, and threats are external conditions that can hinder the organization's ability to achieve its goals.

The present SWOT analysis was the product of a multistage process. The initial impetus was the highlighting of constraints and opportunities for growth within Research–Practice Partnerships (RPPs) at a national symposium organized by the third author in the Faculty. This was followed by a daylong workshop (Faculty Day) attended by faculty members, adjunct teachers, and administrative staff.

Participants in the workshop were divided into groups for thinking about main issues concerning the current and future status of the Faculty. These included issues such as hiring future faculty, preserving and extending tracks, tending to the needs of graduate students, and more. Each group was asked to think about their subject in relation to the SWOT characteristics. The groups then shared their work in the plenary, where additional ideas were raised and discussed.

At the end of this process, a smaller team of senior faculty went through the SWOT table obtained in the workshop, refined entries that needed elaboration, and collapsed redundant entries. The senior faculty then reviewed the final table (see Table 2.2) and agreed on its contents. They made sure that each characteristic listed in Table 2.2 is also related to a specific set of actions that feed into the Faculty's strategic planning.

2.4   Findings

The general theme emerging from the SWOT analysis is as follows: The faculty in its current format has many strengths, including a relatively large number of faculty members and students, compared to other academic institutions; Faculty members are positioned to impact STEM education in Israel; multidisciplinarity facilitates widespread and varied teaching and research collaborations. Weaknesses identified included unstable research budgets and inadequate research support facilities. Opportunities included a growing demand for the Views I and Views II programs and the current need for technology and engineering professionals, driven by a parallel demand for high school teachers in these fields. Finally, threats identified included large proportion (25%) of faculty members who will retire within the next five years relative to the total number of faculty members (3 out of 12).

The action points detailed in the right-hand column of Table 2.2 are an integral part of the SWOT analysis and were key in developing the strategic plan.

2.5   Strategic Plan

This section focuses on four aspects of the Faculty future and strategic plans: (1) hiring plans, (2) teaching load, (3) the Views program, and (4) the graduate program.

Table 2.2 Faculty SWOT analysis

Strengths	
Strengths: faculty and faculty members	
Characteristic	Strategic plan item
The Faculty is the largest STEM education academic unit in Israel in terms of number of faculty members and number of students	Capitalize on the critical mass to maintain excellence and pursue new directions in teaching and research
Tenured, tenure-track faculty members, and adjunct lecturers are professional and dedicated. Each has a proven record of accomplishment in research, teaching, or both	Ensure opportunities for advancement and professional development Enable adequate career opportunities for adjunct staff
The Faculty is well positioned to impact STEM education in Israel:	
Alumni of the various tracks in the Faculty hold positions in which they can influence developments in educational practice and policy through the education system (including Ministry of Education), in industry and the public sector, NGOs, and academia	Maintain high level of interdisciplinarity and multi-sector focus in training of students to develop their awareness of their potential role in education Maintain high level of interaction with education system and adjust teachers' training priorities as appropriate
Faculty members influence developments in educational practice and policy through the dissemination and application of their research, through consulting activities (e.g., with Ministry of Education, Ministry of Science, Technology, and Space, CHE) and by training cohorts of highly skilled educators	
Faculty members are well informed about trends in Israel's education system and able to address programming needs proactively. For example, through the Views programs and through promotion of training of high school teachers with expertise in engineering and technology	
Faculty members, alumni, and Views I/II students have strong ties to industry	Maintain these ties through relevant applied research and the training of educators with industry-relevant skills
Strengths: research	
Characteristic	Strategic plan item
The Faculty's multidisciplinary nature facilitates research collaboration within the Faculty and with other Technion units. Links with the following Technion departments are especially strong: environmental engineering, architecture and town planning, and industrial engineering; electrical engineering, biomedical engineering, and medicine	Actively expand collaborations by engaging with other Technion researchers whose work has synergies with fields in education

(continued)

Table 2.2 (continued)

Strengths	
Research within the Faculty includes important, cutting-edge topics in K-12 and post-secondary STEM education that is relevant within Israel and internationally	Recruit outstanding research student and postdocs with interest in cutting-edge STEM topic Orient recruitment strategy toward hiring new faculty members with appropriate skills and interests Enhance research support infrastructures such as laboratories

Strengths: teaching

Characteristic	Strategic Plan Item
The Faculty program of graduate studies is acknowledged as one of the leading programs in Israel for training high-quality educational researchers and practitioners	Maintain the quality of graduate program; raise and secure funds for graduate scholarships
The Views program is a highly innovative approach to training STEM educators. By recruiting outstanding students who already have strong academic backgrounds in STEM disciplines and in some cases complementary professional experience, the program has increased the supply of excellent high school STEM teachers	Expand opportunities for student recruitment and financial support Recruit new teaching staff in key areas Promote research opportunities for Views II students
The Faculty's teaching activities are closely linked to the school system and education field ensuring both smooth integration of alumni into the teaching profession, maintaining communities of practice and enhancing the application and dissemination of research	Maintain these ties through teacher training and mentoring programs for early career teachers and support through teaching centers
The Faculty is the only unit in Israel to prepare engineering and technology high school teachers	Given current and anticipated future shortages of qualified teachers in these areas must be priorities within Faculty teacher training
Excellence in teaching and pedagogical design evidenced by external recognition; for example, the receipt of the Yanay Award for special excellence in teaching by two faculty members.	

Strengths: impact

Characteristic	Strategic Plan Item
Faculty alumni hold positions in which they can influence developments in educational practice and policy through the education system (including Ministry of Education), in industry and the public sector, NGOs, and academia	Maintain high level of interdisciplinarity and multi-sector focus in training of students to develop their awareness of their potential role in education

(continued)

Table 2.2 (continued)

Strengths	
Faculty members influence developments in educational practice and policy through the dissemination and application of their research, through consulting activities and by training cohorts of highly skilled educators	
Faculty members are well informed about current and future trends in Israel's education system and able to address programming needs proactively. For example, through the Views program and through promotion of training of high school teachers with expertise in engineering and technology	Maintain high level of interaction with education system and adjust teacher-training priorities as appropriate

Weaknesses	

Weaknesses: faculty and faculty members	
Characteristic	Strategic plan item
Changes in education require augmentation of Faculty by researchers having the necessary skill sets	Recruitment of new faculty members is an attempt to respond to these needs and increase the faculty member body and its diversity
We lack faculty members in certain research areas of education and disciplines. Expected faculty retirement in chemistry and engineering education is looming	

Weaknesses: research	
Characteristic	Strategic plan item
Research budgets are not stable	Improve grant writing and increase fundraising efforts
Research support facilities are inadequate	Secure increased funding from the Technion

Weaknesses: teaching	
Characteristic	Strategic plan item
High ratio of students to faculty limits our ability to meet demand for expanded teaching programs and number of mentored research students	Increase number of faculty members through recruitment of excellent Ph.D. students near graduation. We proactively approach such students and encourage them to pursue postdoctoral fellowships
The demographics of faculty members does not represent the diversity of student body and proportion of student from the Arab sector	Make greater efforts to hire new faculty members from underrepresented groups
Lack of well-defined progress track for adjunct lecturers	Introduce formal process for advancement of adjunct lecturers
Mentoring of junior teachers of engineering in the areas of PCK	Prioritize this area within existing mentoring programs

(continued)

Table 2.2 (continued)

Weaknesses	
Weaknesses: finance, administration, and infrastructure	
Characteristic	Strategic plan item
The Faculty building and facilities are aging and in need of major renovations and expansion, including office space, laboratories, conference rooms, future learning spaces (FLS) classrooms, video conference rooms, collaborative learning spaces, library, and equipment	Raise external funds and secure budgets from the Technion for modernization of building and facilities
Insufficient support of management staff, including management of research budgets, purchasing, hosting visits, reserving rooms, and organizing conferences	Clarify job definitions and responsibilities, introduce professional development programs aimed at Faculty management staff
Lack of services and technical staff such as laboratory technicians, statistics, graphics, and language editors	Expand operating budget to meet these needs
Weaknesses: impact	
Characteristic	Strategic plan item
Lack of formalized links between academia and the education system limits the ability to influence. Influence is overly reliant on informal channels and individual initiatives	Maintain networks through outreach infrastructures—teacher centers, etc.
Opportunities	
Characteristic	Strategic plan item
Young and growing Faculty, that is able to draw on its rich prior experience as a department and benefit from advantages accompanying the Faculty status; for example, the opportunity to increase the number of Faculty member appointments	Increase the number of faculty members and integrate early career scholars
Demand among non-Technion graduates for participation in Views I and Views II and potential, dependent on increased funding	Build on early successes of the Views program to recruit new highly qualified students. Expand Technion support and attract financial support from external sources
Informal/lifelong learning has risen in importance within industry, government, and NGO sectors. Links to advanced industry have not yet been fully realized	Engage stakeholders from all sectors, through participation, application, and dissemination of research and by continuing to develop teacher-training courses that are relevant to non-academic environments. Where relevant engage Views students who have professional experience in engaging with these sectors
There are growing needs for technology/engineering professionals accompanied by derived demand for high school teachers in these fields	Increase training opportunities for high school teachers in technology and engineering; hire faculty and adjuncts with appropriate specializations; promote research (especially applied research) in this field

(continued)

Table 2.2 (continued)

Opportunities	
The Faculty and the Technion attract excellent students	Active outreach to prospective students including promotion of STEM education as a field of study and profession
Two education-oriented units within the Technion—the Pre-Academic Unit and the Center for Promotion of Teaching and Learning—have interests and expertise that is complementary to the Faculty's	Promote collaboration with these two units
New York and Chinese campuses offer expanded opportunities for collaboration and for providing students and Faculty members with high-quality international experience	Encourage students and faculty to study and collaborate with these campuses. Develop opportunities for academic exchanges
The relationship of students and faculty members with industry can foster further innovations in the world of educational technologies	The Faculty's teaching programs train professionals with skills to deal with these issues
The global and changing job markets require new teaching and learning methods for interdisciplinary domains and twenty-first century skills	

Threats	
Characteristic	Strategic plan item
Three out of 12 faculty members will retire within the next five years. This is a large proportion of the total number of faculty members. Retirement at the age of 67/68 is mandatory in Israel	Priority must be given to planning for and managing the transition by hiring new faculty members with similar specialties to those of the retiring faculty members and by ensuring that the "institutional expertise" of veteran members is not lost
Heavy reliance on adjunct instructors (relative to the number of permanent faculty members) carries with it the risk that the instructors may not be well integrated	Ensure that processes for integrating of adjuncts are established and ongoing
There is shortage of qualified candidates for recruitment of faculty members in Israel and inability to recruit candidates from abroad due to language barrier	Flexibility in hiring practices, for example, permitting the hiring of promising candidates prior to their gaining international experience and reputation
There is uncertainty regarding the CHE's decisions on the ultimate status of the postgraduate program status	Develop alternate plans for academic programs including support from the Technion. For example, promote Views II as an advanced degree, which will not be affected by the CHE decision
Insufficient support and investment in science and engineering education in government level in Israel	Take part in advancing the issue in policy agendas

2.5.1   *Recruiting New Faculty in the Next Five Years*

Faculty's hiring plans were designed to overcome the Weakness of lack of faculty in specific areas, the Threats of expected retirement and to take advantage of the Opportunities of young and growing faculty. As such, they are aligned with the demographics of its members, with developments in the field of STEM education as well as with changes in the student population and the needs of the education system. Particular attention is paid to ensuring that all STEM fields continue to be represented, and that our unique approach to interdisciplinary teaching and research is maintained.

The Faculty's demographic differs from the "classic" academic model. Many of our researchers and lecturers completed their Ph.D.'s and joined the Faculty after a number of years of professional experience within the education system as teachers. As a result, the age at entry is higher than in many other academic units. This has several implications, such as the average age within the Faculty is older. A similar challenge relates to the problem of finding faculty member candidates who have the requisite postdoctoral experience abroad prior to joining the faculty. As a result, an important element in the Faculty's recruitment strategy is to consider recent Ph.D. graduates and some who are very close to completion and who have published at least one or two peer-reviewed papers, but who still lack international experience. These young scholars are hired at a Lecturer rank, which is the lowest rank in the tenure track (similar to the British system) and are expected to obtain extended international experience as postdocs or visiting scholars during their tenure-track period. The faculty had good experience with such a process when previously hiring biology education and math education faculty members.

Given the anticipated retirement of several faculty members over the next five years, criteria for assessing candidates for recruitment therefore include demonstrated excellence in STEM education, especially at the high school level as well as DBER in higher education as a primary research interest. This is particularly true of technology and engineering education since the Technion is the only university that trains high school engineering teachers in Israel. Cross-disciplinary educational research experience is also a highly desirable qualification. This includes, experience in educational policy, assessment, advanced information and communication technologies, informal learning, science communication, creativity, giftedness, learning disabilities, and game-based learning.

Recruitment also focuses on the candidates' potential to develop high-level collaborations with other units within the Technion, within Israel and abroad (Dori and Belcher 2005). Similarly, the capacity to train future teachers in transformative education models such as MERge (Hazzan and Lis-Hacohen 2016) that promotes the amalgamation of management, education, and research is highly valued.

2.5.2   Teaching Load

As a general rule, each faculty member teaches two courses per semester, totaling four to six weekly hours. These courses usually include one undergraduate and one postgraduate course. Some faculty members also take academic responsibility for additional courses that are taught by adjunct faculty. In addition, adjuncts are an integral part of our teaching staff. Most adjuncts also maintain in roles in schools and colleges of education (teacher training institutions). Thus, the close relationship with them maintains the Strength of ties with the educational system.

With the introduction of the Views program, the Faculty's student population has grown dramatically, and this growth is expected to continue. Therefore, and in line with our goal to overcome the Threat of reliance on non-tenured adjuncts, recruiting and sustaining adjunct faculty as teaching faculty is a high priority. Recruitment targets Ph.D. alumni of the faculty and others from industry, government, and other sectors with expertise in particular fields.

2.5.3   The Views and Graduate Programs

As mentioned earlier, the Views program is targeted at excellent B.Sc. graduates of the Technion, training them in the course of around 1–2 days a week over 4–6 semesters to become high school teachers. Views I offers an additional minor in STEM education, which includes a teaching certificate, while Views II offers an M. Sc. degree. Both programs provide students with full tuition scholarships. The two Views programs have become the Faculty's flagship teacher education programs and figure highly in the Strengths rubric of the SWOT analysis. Therefore, consolidating Views I and advancing Views II is a major aspect of the Faculty's strategic plan. So is developing and disseminating research on Views as a model for STEM teacher training, as well as its impact on STEM education within the school system and Israeli society as a whole. The first five years of Views I have shown that the program's alumni have skills in management, teamwork, and negotiation as well as the ability to cope with vagueness and manage risks. These are all highly relevant for mediating and applying STEM in multiple sectors and over the life course.

Views I graduates teach in Israeli high schools and have included over the past 5 years approximately 350 students. To overcome the Weakness of Views programs being targeted only at Technion graduates, the strategic plan includes raising funds to recruit outstanding graduates of other Israeli universities. This includes fundraising to provide scholarships that would support these candidates in a manner similar to that provided to the Technion alumni (currently, non-Technion degree holders are ineligible for the tuition support and can get partial support from the Ministry of Education).

The strategic plan for the Views program includes an extension in two directions—one is high school management and the other is educational technologies. These are elaborated below.

As mentioned in the Strengths component of the analysis, the Faculty program of graduate studies is acknowledged as one of the leading programs in Israel for training high-quality educational researchers and practitioners. The Faculty devotes thought and resources to continuous upgrading the program to ensure excellence in research and educational leadership. The recent recognition of our department as a faculty demands further development of the graduate program. Directions for the next five years include:

To take into account the Opportunity of Technion's international campuses, the strategic plan includes internationalization of the program. This is in line with the Technion's emphasis on the development of international collaborations in research and education. As part of this effort, the Faculty plans to open its gates to international graduate students in STEM education research. The Technion campuses in New York and China will also provide a platform for the effort. Challenges involve finding the appropriate balance of Hebrew and English languages, inclusion of students with diverse backgrounds, and the fact that they will need to function in the school system in which the language spoken is Hebrew or Arabic.

Excellence programs: In view of the Threat of shortage in young and promising educational researchers and/or leaders in the educational system, the Faculty plans to establish (a) flexibility in hiring practices for tenure-track positions, (b) a special framework for outstanding graduate students interested in academic careers, and (c) a leadership program for leaders in the educational system, preparing the next generation of school principals or national STEM superintendents. These last two programs will offer intensive courses, close involvement in research, and international student exchange.

Advanced educational technology infrastructure: To overcome the Weakness of the Faculty's facilities aging and in need of major renovation, the strategic plan includes raising and securing funds for such renovations. Modern STEM education is based on experiential learning in technology-rich environments. The Faculty believes that such infrastructure is essential for the professional development of graduate students as prospective teachers and researchers. As a first step, we are currently renovating the library and redesigning it into a collaborative learning space.

Interdisciplinary Graduate Study Programs: In line with the Opportunity of global and changing job markets, that require new teaching and learning methods for interdisciplinary domains, we plan to develop two Interdisciplinary Graduate Study Programs: (a) in sustainability, partnering with environmental engineering, architecture and town planning, and industrial engineering; and (b) in mind, brain, and education, partnering with several engineering units, such as biomedicine engineering, industrial engineering, as well as medicine.

2.6 Summary

In this chapter, we provided a detailed case study and SWOT analysis of the Faculty of Education in Science and Technology at the Technion as a case in point. It demonstrates both the viability of this academic unit and the challenge it faces to maintain and strengthen its position as a world-class unit of STEM education. Through the lens of this particular academic unit, we have been exposed to the problems and difficulties similar academic units or departments worldwide face as they struggle to deliver STEM education of the highest quality with limited resources. Projecting the specific situation of the Faculty to other similar academic units worldwide can serve as a yardstick to compare and contrast the challenges others in the domain of STEM education face as they struggle to fulfill their mission. Specifically, this monograph might provide assistance to those who are uncertain about the desired structure of newly created STEM academic units with respect to whether they should be distributed across the disciplinary STEM units or incorporated in the same unit. The analysis here clearly points to the benefits of having the STEM educators in an integrated unit while interacting with the disciplinary scientists about education.

References

Afzalur Rahim, M. (2002). Toward a theory of managing organizational conflict. *International Journal of Conflict Management, 13*(3), 206–235.

Barak, M., Harward, J., Kocur, G., & Lerman, S. (2007). Transforming an introductory programming course: From lectures to active learning via wireless laptops. *Journal of Science Education and Technology, 16*(4), 325–336.

Becher, T., & Trowler, P. (2001). *Academic tribes and territories: Intellectual enquiry and the culture of disciplines*. UK: McGraw-Hill Education.

Biglan, A. (1973). The characteristics of subject matter in different academic areas. *Journal of Applied Psychology, 57*(3), 204–213.

Coppola, B. P., & Krajcik, J. S. (2014). Discipline-centered post-secondary science education research: Distinctive targets, challenges and opportunities. *Journal of Research in Science Teaching, 51*(6), 679–693.

Dori, Y. J. (2007). Educational reform at MIT: Advancing and evaluating technology-based projects on- and off-campus. *Journal of Science Education and Technology, 16*(4), 279–281.

Dori, Y. J. (2008). Reusable and sustainable science and engineering education. *Journal of Science Education and Technology, 17*(2), 121–123.

Dori, Y. J., & Belcher, J. W. (2005). How does technology-enabled active learning affect students' understanding of scientific concepts? *The Journal of the Learning Sciences, 14*(2), 243–279.

Dori, Y. J., Hult, E., Breslow, L., & Belcher, J. W. (2007). How much have they retained? Making unseen concepts seen in a freshman electromagnetism course at MIT. *Journal of Science Education and Technology, 16*(4), 299–323.

Hativa, N., & Marincovich, M. (1995). Editors' notes. *New Directions for Teaching and Learning, 1995*(64), 1–4.

Handelsman, J., Ebert-May, D., Beichner, R., Bruns, P., Chang, A., DeHaan, R., et al. (2004). Scientific teaching. *Science, 304*(5670), 521–522.

Hazzan, O., Lis-Hacohen, R. (2016). The MERge Model for Business Development: The Amalgamation of Management, Education and Research, SpringerBriefs in Business. http://www.springer.com/us/book/9783319302249

Lipson, A., Epstein, A. E., Bras, R., & Hodges, K. (2007). Students' perceptions of Terrascope, a project-based freshman learning community. *Journal of Science Education and Technology, 16*(4), 349–364.

Marbach-Ad, G., McAdams, K. C., Benson, S., Briken, V., Cathcart, L., et al. (2010). A model for using a concept inventory as a tool for students' assessment and faculty professional development. *CBE-Life Sciences Education, 9*(4), 408–416.

Mintz, K., Talesnick, M., Amadei, B., & Tal, T. (2013). Integrating sustainable development into a service-learning engineering course. *Journal of Professional Issues in Engineering Education and Practice, 140*(1), 05013001.

Mitchell, R., Dori, Y. J., & Kuldell, N.H. (2011). Experiential engineering through iGEM – an undergraduate summer competition in synthetic biology. *Journal of Science Education and Technology, 20*(2), 156–160.

Singer, S. R., Nielsen, N. R., & Schweingruber, H. A. (2012). *Discipline-based education research*. Washington, D.C.: The National Academies.

Shulman, L. S. (1986). Those who understand: Knowledge growth in teaching. *Educational Researcher, 15*(2), 4–14.

Tsaushu, M., Tal, T., Sagy, O., Kali, Y., Gepstein, S., & Zilberstein, D. (2012). Peer learning and support of technology in an undergraduate biology course to enhance deep learning. *CBE-Life Sciences Education, 11*(4), 402–412.

Chapter 3
Research–Practice Partnerships in STEM Education: An Organizational Perspective

Orit Hazzan, Einat Heyd-Metzuyanim & Yehudit Judy Dori

Abstract The research presented in this chapter examines Research–Practice Partnerships (RPPs) in Israeli Science, Technology, Engineering, and Mathematics (STEM) education. We performed a SWOT analysis, where we categorized factors into Strengths, Weaknesses, Opportunities and Threats on data collected before, during, and after a conference that was devised to examine the state of STEM RPPs in Israel. The overall data analysis revealed the following theme: Studies in STEM education focus on what goes on in the schools and the need for RPPs. Yet, RPPs face obstacles rooted in the organizational structure and culture of the two RPP partners: the Research—Academia in Israel; and the Practice—Ministry of Education (MoE). Therefore, while both the education system and academia agree on the necessity to collaborate, these collaborations are not fully actualized, and RPPs in STEM education in Israel do not invest in the most critical problems, the investigation of which is crucial for the economic growth and development of the state of Israel. The SWOT analysis presented in this chapter, which is based on data gathered from representatives of all STEM education sectors, enabled us to deepen our understanding of how this situation can be improved: specifically, (1) leveraging the multiple activities in STEM education that exist in Israel, which was identified as the most meaningful **strength**; (2) bridging the cultural gap between academia and the educational system, identified as the most meaningful **weakness**; (3) capitalizing on the increased attention and importance attributed to STEM education in the country, identified as the most meaningful **opportunity**; (4) overcoming the frequent policy changes that take place in the MoE due to political forces in Israel, which are beyond the control of the educational system, identified as the most meaningful **threat**; (5) we also found that academia is perceived as the stakeholder responsible for promoting RPPs in STEM education in Israel.

Keywords Research–Practice Partnerships · STEM education · Organizational perspective · SWOT analysis · Academia · Ministry of education

3.1 Introduction

The research presented in this chapter focuses on cross-sectoral research-based collaboration in Science, Technology, Engineering and Mathematics (STEM) education in Israel. The concept of cross-sectoral collaborations refers to partnerships among organizations from three sectors: the first—government and local authorities; the second—for-profit corporations; and the third—nonprofits and NGOs (public universities and other non-governmental organizations). As it turns out, even though many collaborations among the different sectors related to STEM education exist in Israel, they are mostly neither research-based nor methodologically documented (with some exceptions, such as Avargil et al. 2013; Zohar in press; Zohar and Cohen 2016). Therefore, the knowledge generated as a result of the collaboration is not explored and does not serve other educational initiatives. In addition, organizations, in general, and the Ministry of Education (MoE), in particular, are not exposed to the different possible types of productive collaboration that may leverage the Israeli education system in general and STEM education in Israel in particular. Since we focus on cross-sectoral research-based collaborations, we will use the common term Research–Practice Partnerships (RPPs) (Coburn and Penuel 2016), which has originated from the idea of collaboration between teachers and researchers.

The chapter is organized as follows: Sect. 3.2 presents the research background, and Sect. 3.3 presents the research structure. The data analysis, presented in Sect. 3.4, is organized according to the SWOT analysis components: Strengths, Weaknesses, Opportunities and Threats. Based on the data analysis, and as is commonly done with SWOT analysis, Sect. 3.5 lays out several recommendations that can promote RPPs in STEM education in Israel. In Sect. 3.6, we conclude our chapter.

3.2 Background

RPPs often take the shape of a linear model (Stein and Coburn 2010). According to this model, research results usually flow in one direction—from basic research, through applied research and implementation stages, to the dissemination stage. Penuel and his colleagues (Penuel et al. 2015) highlight the illusion underlying the "linear model":

> Researchers often imagine that the best way to bridge that (research-practice) gap is to translate basic research on learning into interventions that are feasible for teachers to implement, effective for a wide range of students, and accessible to any student who might benefit from them (p. 182).

Yet despite the seemingly attractive vision of neatly "packaged" intervention programs coming out of the researchers' laboratory and "translated" into useful wide-scale educational programs, the situation in Research–Practice Partnerships is

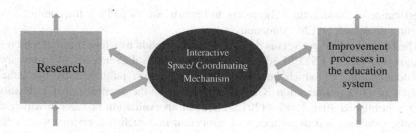

Fig. 3.1 Dual Model for RPPs reproduced from Stein and Coburn (2010, p. 8)

much more complex. Specifically, Penuel et al. (2015) claim that the linear or pipeline model has three limitations:

(1) It does not encourage a relationship of mutualism and reciprocity between researchers and practitioners; (2) it does not fit situations where interventions move from one setting to another; (3) it conveys a narrow use of research as solely a form of providing evidence for decision making, whereas successful RPPs "practitioners can, and do, value research that helps them gain new insights into problems and that facilitates the search for new kinds of solutions to persistent problems" (p. 186).

Stein and Coburn (2010), in examining a series of RPPs in the 1990s and 2000s in the US education field, offered an alternative to the linear model. According to their dual model, information between researchers and practitioners flows in both directions (see Fig. 3.1) through an "interactive space" where researchers and practitioners regularly meet, work together, and exchange information.

Despite the rise in RPPs in the USA over the last two decades and the conceptual work that has been done to characterize them (e.g., Coburn and Stein 2010; Penuel et al. 2015), not much is known about the enablers and constraints that allow forming successful partnerships. In a recent review, Coburn and Penuel (2016) address Research–Practice Partnerships in Education: Outcomes, Dynamics, and Open Questions. They argue that "although there is evidence of success of the interventions developed within RPPs in other fields, research on the impact of RPPs in education is sparse and focused on a narrow range of outcomes." Coburn and Penuel (2016) also claim that "Most research on the outcomes of RPPs in education and other fields has focused on the impact of interventions developed in the context of a partnership. Thus, they do not investigate the impact of the partnership itself or other outcomes of RPPs" (p. 49). Accordingly, they propose a research agenda for investigating RPPs that includes four areas of focus: outcomes, comparative studies, strategies, and political dimensions.

Our interest in this study lies specifically in strategies, or on how to promote RPPs and their outcomes. Coburn and Penuel (2016) called for studies focusing on "the relative strengths and weaknesses of strategies for addressing such common challenges as persistent turnover, the need to create shared language or work practices, fostering trust, and ways to work with and across multiple levels of educational systems." (p. 52). We take a wider approach and expand this investigation to the structural enhancers and constraints of RPPs at the country level focusing on Israeli STEM education.

The concepts of linear versus dual model will provide us a lens through which to examine the RPPs in STEM education in Israel. Several events led us to research this domain, one of which being an evaluation report published by the Israeli Council of Higher Education (CHE, Committee for the Evaluation of Education 2015[1]), published after 3 years (2012–2014) of an evaluation process in which all schools, faculties, and departments of education in Israeli universities were evaluated by an international committee assigned by the CHE. In its general report (CHE 2015), the committee wrote:

> Research is the lifeblood of all academic units of a university, but professional schools have a social obligation that goes beyond conducting research for research's sake: they have a duty to improve the society of which they are a part. For an education faculty, this means that educational research must find its way into the hands of the teachers, students, principals, parents, and supervisors who staff and run the nation's educational institutions (p. 8).

> Investigating [issues that are critical for the educational system....].... are the ones least studied by its scholars (p. 12).

Being aware of possible models for RPPs (e.g., Stein and Coburn 2010), we decided to examine what kinds of RPPs and of what scope are carried out in STEM education in Israel, as well as barriers they face and opportunities these RPPs can exhaust. In practice, to improve our understanding of the above-mentioned issues (mainly, strategies) and to expose the roots of the CHE report's criticism, we adopted a research–practice approach inspired by RPPs. Specifically, we gathered data through a conference together with additional survey data from all sectors in Israel who have interest in STEM education (see Sect. 3.3—Research Structure).

To facilitate the examination of strategic dimensions of RPPs in Israel, specifically, what targets they achieve, do they exhaust their potential, and how they are managed, we applied the SWOT analytical tool. SWOT is an acronym of Strengths, Weaknesses, Opportunities, and Threats. SWOT analysis is one among many 2 × 2 matrices (Lowy and Hood 2010) organizations use to manage their business (either financial or social), make decisions, and define strategies. It enables examining strategies that amplify the strengths and the opportunities and thus defending the organization from its weaknesses and the environment threats (Barney 1995). A SWOT analysis is applied for companies, products, places, etc. In short, Strengths refer to characteristics of the analyzed object that give it an advantage over others; similarly; Weaknesses refer to characteristics that place the analyzed object at a disadvantage relative to others; Opportunities are phenomena, in most cases outside the organization, which the analyzed object/organization can take advantage of and exploit for the advantage of its goals; Threats also refer to elements in the surrounding environment of the organization, but, unlike opportunities, threats can hinder the organization from achieving its goals.

---

[1]The third author was a member of the CHE committee.

SWOT analysis is used as a tool for organizational analysis also for public organizations, such as schools and hospitals (Rego and Nunes 2010). For example, a SWOT analysis was carried out for the evaluation of educational initiatives, such as the integration of information technologies (Sabbaghi and Vaidyanathan 2004), and for defining a risk management plan for STEM education in Israel (Even-Zahav 2016). The first author of this chapter applied this analysis framework for the analysis of Israel software start-up ecosystems (Kon et al. 2014) and for the analysis of change processes in higher education (Hazzan and Lis-Hacohen 2016).

We propose that RPPs in STEM education in Israel is an important topic for investigation for two main reasons. First, due to the significant contribution of STEM subjects to Israel's economic growth and development, STEM topics attract the attention of all sectors in Israel, and therefore, RPPs are in the interest of all sectors. Second, Israel is a small country, with about 8.5 million citizens (more or less equal to New York City population). This size enables to reveal almost a full picture of the current situation in Israel with respect to RPPs in STEM education and to draw conclusions from which larger countries can learn. A similar argument has led us to use programs, such as the *Views* program at the Technion, which trains Technion graduates in science and engineering for teaching careers, as an example case of an effort to mobilize highly qualified engineers and scientists into STEM education (http://cgi.stanford.edu/~dept-ctl/cgi-bin/tomprof/enewsletter.php? msgno=1337). Others have similarly used Israel as a field of piloting innovative ideas (e.g., Shai Agasi's electric car case told in Sensor and Singer's book Start-up Nation, 2009).

3.3 Research Structure

In this section, we present the research objectives, questions, setting, population, and data gathering tools.[2]

Research objectives:

(a) To describe the scene of RPPs in STEM education in Israel as well as its strengths, weaknesses, opportunities, and threats of such partnerships.
(b) Based on (a), to formulate recommendations whose aim is to improve information flow in two directions—from the education system to the academia and vice versa.

From these research objectives, the following research questions were derived.

---

[2]The Technion—Israel Institute of Technology—Committee for Ethics in Social Sciences—approved the research.

Research questions

- What factors (strengths) promote RPPs in STEM education?
- What obstacles (weaknesses) do RPPs face, in both the education system and the academia?
- What opportunities can RPPs in STEM education in Israel exploit?
- What threats do RPPs in STEM education face?
- In what ways can RPPs be improved?

Research setting and participants

Our research populations included a part of the Israeli community of STEM education—researchers and partners from all sectors—which were gathered in our Faculty for a one-day conference that took place on May 24, 2016. The title of the conference was *Cross Sectoral Research Based Cooperation in STEM Education: The Case of the Israeli Education System*. The conference agenda is presented in Table 3.1. In addition, in the validation stage of our findings, the research

Table 3.1 Conference agenda[a]

9:00–9:30 Assembling and coffee
9:30–9:45 Introductory words: • Prof. Peretz Lavie, President of the Technion   • Prof. Yehudit Judy Dori, Dean of the Faculty of Education in Science and Technology, Technion
9:45–10:00 Dr. Einat Heyd-Metzuyanim: *The main questions for discussion in the conference*
10:00–11:00 Prof. Mary-Kay Stein, University of Pittsburgh: Plenary Talk
11:00–11:30 Reactions[b]:   • Prof. Bat-Sheva Eylon, Former Head of the Department of Science Teaching, Weizmann Institute of Science   • Dr. Itay Asher, Acting Chief Scientist, Ministry of Education   • Prof. Anat Zohar, Former Head of the Israel MoE's Pedagogical Council, Hebrew University of Jerusalem
11:30–11:45 Coffee break
11:45–12:45 Researchers' panel: *What is the potential contribution of studies in STEM education to the field of education?*
12:45–13:30 Lunch
13:30–14:15 Ministry of Education, industry, NGOs and foundations' panel: *What does the educational system need from academic research?*
14:15–15:15 Roundtables of the conference participants   – Presentation of research projects by faculty members, postdocs, and doctoral students   – Discussion concerning the main questions of the conference
15:15–15:45 Summary of the roundtables by a representative from each table
15:45–16:00 Summary and farewell—Prof. Orit Hazzan (via Skype)

[a]We open the Findings section of this paper with an illustration of how the messages delivered by the plenary speakers of the workshop can be naturally analyzed by the SWOT framework of RPPs in STEM education in Israel
[b]Scholar's details appear here with their permission

population was expanded and included additional stakeholders, such as executives and senior officers in a variety of organizations that deal with STEM education in Israel—in the public (e.g., MoE), for-profit and nonprofit sectors.

One hundred and five (105) participants took part in the conference: 62 were from the academia (59%), 26 from the education system (25%), 4 from the industry (4%), and 13 from NGOs and philanthropic organizations (12%).

Data gathering tools

Data were collected prior, during and after the conference, by the following research tools.

Pre- and post-surveys

Online surveys were distributed to all conference participants who registered to the conference. Responding to the survey was optional and anonymous.

In the pre-conference survey, the research rationale was described and data were collected about the participants' experience with RPPs. Fifty-eight (58) participants answered the questionnaire: 35 (60.3%) from the academia, 14 (24.1%) from the education system, 4 (6.9%) from philanthropic organizations, 2 (3.4%) from the industry, and 3 (5.2%) indicated 'other'.

In the post-conference survey,[3] participants were asked to reflect on their experience in the conference, to address their future intention to be involved in RPPs, and to suggest recommendations for the promotion of RPPs in STEM education in Israel. Forty (40) participants answered this survey: 32 (80%) from the academia, 4 (10%) from the education system, 1 (2.5%) from philanthropic organizations, and 3 (7.5%) indicated 'other'.

Submissions for presentation in the roundtables in the form of project abstracts

The conference participants were invited to submit a description of research works relevant for the conference themes—RPPs in STEM education in Israel. In addition to the project description, submission included the following topics for discussion in the conference:

- Potential contribution of the project/research to Israeli STEM education.
- The project/research relevance for academic research.
- Opportunities and barriers faced in collaborative work with the education system.

Twenty-seven (27) research descriptions were submitted and analyzed. The submissions were categorized into five topics: information and communication technologies in STEM education, management of educational systems, math education, social aspects of STEM education and teaching skills and methods. These research studies were discussed in the roundtable session of the conference. The five

---

[3]The post-conference survey is in Hebrew.

roundtables were facilitated and headed by faculty members from our Faculty according to their expertise. Each roundtable included about 20 participants from all sectors: 5–6 presenters, 6–8 researchers from the academia, 5–6 practitioners from the education system, and 3–4 representatives of the industry and the NGO non-profit sector.

Observations in the conference

The conference sessions were video recorded and analyzed according to the following themes:

- Factors/strengths which promote RPPs in Israel.
- Obstacles/weaknesses the participants face while engaged in RPPs.
- Opportunities for RPPS that can be actualized to promote STEM education in Israel.
- Threats that should be considered when RPPs are developed.
- Other ad hoc themes.

Validation survey[4] of the SWOT analysis

Following the data analysis by the SWOT framework, a survey[5] was distributed that aimed to confirm the data analysis, in particular, the categorization of the different factors of the RPPs according to the components of the SWOT analysis and the proposed recommendations.

The validation survey underwent two phases of validation. In the first phase, we distributed an online questionnaire (Validation Questionnaire 1) to 10 respondents from varied backgrounds (doctoral students, teachers, and NGOs). The questionnaire listed in a random order 14 factors that were identified as either strengths, weaknesses, opportunities, or threats. The responders were requested to categorize these factors as either a strength, a weakness, an opportunity, or a threat with respect to RPPs. The reactions to this questionnaire were unanimous: Almost all respondents declared that the questionnaire is confusing and that they were unable to understand the factors and to categorize them as was asked. In response, we distributed a second version of the validation questionnaire (Validation Questionnaire 2).

In the Validation Questionnaire 2, we asked the respondents to indicate the extent to which they agree with our categorization according to the SWOT components and to offer an alternative categorization, if they felt the factor belongs to another SWOT category. In addition, respondents were asked for each category to

---

[4]In qualitative research, this activity would have been called credibility, which is the associate term to internal validity used in quantitative research. The credibility criteria are said to insure that the results of qualitative research are credible from the perspective of the research participants. However, we decided to use the term validation survey for two reasons: First, it is the more common term which reflects our intention; second, this phase mixed both qualitative and quantitative methods.

[5]In Hebrew.

indicate the factors most and least significant to the success of RPPs in Israel. Finally, they were asked to rank the agreement level with respect to each of our proposed recommendations for the promotion of STEM education in Israel and to suggest additional ways by which RPPs in STEM education can be fostered and improved.

Since in the conference itself the representation of the different sectors was not balanced, the survey was distributed also to a group of seniors in different for-profits and nonprofits organizations who are involved in STEM education. Forty-one responses were received, from which only 18 participated in the conference (3 of the 18 only partially), with the following distribution between sectors: 40% ($n = 17$) from the academia and 60% ($n = 25$) from non-academic institutions (one belongs to two sectors). The 60% from the non-academic institutions were distributed as follows: 24.4% ($n = 11$) from NGOs, 19.5% ($n = 9$) from the education system, and 17.1% ($n = 5$) from the industry.

The responses to the validation survey were found to be reliable by two indicators. First, relatively high agreement level was expressed with respect to whether a specific factor had been categorized properly in the SWOT analysis. Second, consistency was observed between the agreement level and the indication of the most significant factor in each category.

We illustrate these arguments with respect to the identified strengths (see Table 3.4 in Sect. 3.4). The strength "multiple activities in STEM education" got the highest level of agreement among the three identified strengths with respect to whether it indicates a strength, and at the same time, was selected by 47% of the responders as the most significant strength (the other two strengths were indicated as the most significant strength by 32% and 21% of the participants).

In addition, when relevant, we address similarities and differences between sectors, with respect to their choice of the most significant factor in each component of the SWOT analysis. In the case of strengths, for example, we see that the importance the academia attributes to the strength "large research community" is larger than the importance the other sectors attribute to this strength.

Research participants

Table 3.2 presents the representation of the different sectors in the data collected by the different research tools. Since several participants identified themselves as belonging both to the academia and to another sector, the sum of column is sometimes bigger than the total number of participants.

3.4 SWOT Analysis of RPPs in STEM Education in Israel

In this section, we present our SWOT analysis of the RPPs in STEM education in Israel. The data analysis revealed that currently, the infrastructure for RPPs in STEM education exists, as expressed in the Strengths and Opportunities. However, at the same time, these RPPs do not exploit their potential due to cultural gaps

Table 3.2 Summary of data gathering tools and population

Group		Pre-survey (n = 58)	Conference (n = 105)	Post-survey (n = 40)	Validation survey (n = 41)
Academia—The *Research* of RPPs	**Academia**	**37 (2 also in the education system)**	62	**34 (2 also in the education system)**	**17 (1 also in the industry)**
Not Academia— The *Practice* of RPPs	**Total— non-academia**	23	43	8	25
	Education system	16 (2 also in the academia)	26	6 (2 also in the academia)	9
	Industry	2	4	–	5 (1 also in the academia)
	Nonprofit	4	13	1	11
	Other	1	–	1	–

between academia, the education system, organizational structures, and political processes, expressed as Weaknesses and Threats.

We first examine the main messages delivered by the plenary speakers in relation to strengths, weaknesses, opportunities, and threats in STEM education in Israel (Table 3.3). We place special attention to these perspectives, due to the speakers' accumulative experience and central roles in STEM education both in the MoE and in the academia.

In what follows, for each component of the SWOT analysis, we first describe the factors associated with it and illustrate the analysis with relevant data. Table 3.4 presents the ranking of the factors associated with said component, as was expressed in the validation stage.

3.4.1   Strengths of RPPs in STEM Education in Israel

Three major strengths of RPPs in STEM education were identified based on the data analysis: (a) multiple activities in STEM education, (b) a large research community in higher education departments in the universities that specialize in STEM subjects, and (c) other research institutions that engage in STEM education research and development. In what follows, we elaborate on each strength.

(a) Multiple activities in STEM education

This strength refers to the multitude of activities happening at this time in Israel that are either specifically tailored for enhancing STEM education or significantly impact it. One example of such projects is the "5 times 2" project, an initiative of

Table 3.3 Illustration of the components of the SWOT analysis as expressed by the plenary speakers of the conference

Strengths	Weaknesses
• The multiple activities in STEM education that currently exist in Israel, which was identified as the most significant strength in our study: *Research-based collaborations in the areas of STEM education exist in Israel for many years, and Israel had highly significant achievements in this area* • Large research communities in the different STEM subjects in academic departments: *We do continue studying problems that arise from the field, although with funds provided mainly through scholarships that the Technion provides to Ph.D. and Master's students, rather than funds provided by Practice partners*	• The cultural gap existing between the academia and the education system: *The discourse between the academic world and the practical world has been changing in the world. I am afraid that in Israel, we have not seen yet a sufficient change, especially with respect to all issues related to the relevance of research to the education field The academia is interested in the creation of new knowledge, in a deep and slow process. The Ministry of Education is eager for much quicker synthesis of existing knowledge, for both the support of policy formulation and the improvement of the practice of school principals and teachers. The office of the Chief Scientist at the MoE tries to promote these two targets* • Academic culture and structure: *while we are dealing with the promotion of research-based collaboration between the academia and the education system, we face several obstacles which prevent these collaborations. In many cases, selective funds, such as ISF (Israel Science Foundation), reject a research proposal only because it is "too practical." On the other hand, the Center of Science Education (MALAM) not only funds a small-number of projects, but also emphasizes only development and does not fund research works which accompany the development of curriculum, but rather, it funds only limited feedback on the developed curricular materials. This situation must be changed to allow us to conduct research that directly impacts the education system*
Opportunities	**Threats**
• The cross-sectoral—that is, beyond first sector—awareness to STEM education in Israel: *There are technological developments which are expected to influence the education system in the next decades and that we should pay attention to both in the education system and at the Technion* *The involvement of the third sector and the industry creates new opportunities*	• Decrease in government funds some of which may result from transferring the responsibility for the education system from the state to private organizations: *recent developments, such as the reduced scope of activity of the Center for Science Education (MALAM), led to the reduction of research-based collaborations* • Priority change due to political circumstances: *an additional problem is the frequent change of Ministers of Education which leads to strategy changes. These changes make it difficult for the education system to progress steadily toward clear targets*

Table 3.4 Summary of validity scores for SWOT analysis

Strengths	Strength 1	Strength 2	Strength 3	
	Multiple activities in STEM education	Large research communities	Additional Israeli research institutions in STEM education	
Level of agreement with the factor being a strength (on a scale 1–3) $N = 41$	$N = 39$ Average 2.38	$N = 33$ Average 2.24	$N = 38$ Average 2.05	
Did not know	2	8	3	
Selected as the most significant strength ($N = 34$)	16	11	7	
Weakness	**Weakness 1**	**Weakness 2**	**Weakness 3**	
	Cultural gap	Academic culture and structure	Regulations in the MoE[a] Average = 2.29 (average of averages of the two items)	
Level of agreement with the factor being a weakness (on a scale 1-3) $N = 41$	$N = 41$ Average 2.63	$N = 37$ Average 2.32	Lack of incentives for teachers and principals $N = 38$ Average 2.55	Research approval in the MoE $N = 32$ Average 2.00
Did not know	0	4	3	9[b]
Selected as the most significant weakness ($N = 22$)	13	4	4	1
Opportunities	**Opportunity 1**		**Opportunity 2**	
	Cross-sectoral—beyond first sector—awareness due to		CHI evaluation report	
	Reduction in Israel's results in international tests	Shortage in hi-tech		
Level of agreement with the factor being an opportunity (on a scale 1–3) $N = 41$	$N = 38$ Average 2.16	$N = 39$ Average 2.15	$N = 32$ Average 2.00	
Did not know	3	3	2	9
Selected as the most significant opportunity ($N = 29$)	10	15	4	
Threats	**Threat 1**		**Threat 2**	
	Priorities change		Responsibility is transferred from the state to private organizations	

(continued)

Table 3.4 (continued)

Level of agreement with the factor being a threat (on a scale 1–3) $N = 41$	$N = 37$ Average 2.70	$N = 35$ Average 1.97
Did not know	4	6
Selected as the most significant threat ($N = 31$)	26	5

[a]The "regulation in the MoE" weakness was formulated first as two weaknesses: "the Chief Scientific Office regulatory operations" and "lack of incentives to principals and teachers to participate in research". We have decided to unify them in discussion in this paper to highlight the fact that regulations and cultural elements, both in the academia and in the education system, contribute to the significant weakness of a cultural gap existing between the two systems
[b]With respect to the "regulations in the MoE," we mention the relatively high number (9) of responders who did not know what it means

the MoE aimed at doubling the number of Israeli high school graduates majoring in mathematics, science, and technology.[6] This project was presented and mentioned by several of the participants in the MoE, industry, NGOs, and foundations' panel. Organizations from all sectors are involved in the promotion of this initiative and collaborate in the framework of the Collective Impact model (Kania and Kramer 2011). Nevertheless, the documentation of this initiative (see http://www.5p2.org.il/model-of-operation/) also indicates that its primary focus and financial support are placed on the practice aspect, while the research aspect, and its associated RPPs, are not inherently promoted and viewed as crucial elements that can support and enhance the project success.

In addition to projects that were specifically mentioned in the plenary panels and in the roundtables, the high involvement of conference participants in RPPs was identified from the pre-questionnaire, in which they were asked to indicate in how many RPPs they have participated. Among the 58 participants who answered this question, nine (15.5%) participated in at least eight RPPs, two (3.4%) participated in 5–7 RPPs, eighteen (31%) participated in 2–4 RPPs, twelve (20.7%) participated in one RPP, and seventeen (29.3%) did not participate in any RPP. Strengths (b) and (c) presented below elaborate on the research frameworks of these RPPs.

(b) **Large research communities in the different STEM subjects in higher education departments**

This strength refers to the exceptionally large research community in Israel devoted to STEM education in all STEM subjects. This research community includes one faculty (at the Technion) and one department (Weizmann Institute) solely devoted to science and mathematics education. For example, the Faculty of Education in Science and Technology at the Technion has faculty members that span all of the STEM education fields: mathematics, physics, chemistry, biology, computer

---

[6]See http://www.5p2.org.il/english/.

science, environmental sciences, and two technological education tracks, mechanical Engineering and electrical Engineering. Other universities, such as Ben-Gurion University, Tel Aviv University, Bar-Ilan, and Haifa University, have units, programs, or departments devoted to STEM education or to mathematics education.

The proliferation of research activity in the field of STEM education can be seen in the large representation of Israeli scholars in international conferences such as National Association for Research in Science Teaching (NARST), International Group for the Psychology of Mathematics Education (IGPME), and Special Interest Group of Computer Science Education (SIGCSE). As an example for the magnitude of research communities in comparison with other research communities around the world, we present two examples: In the International Group for the Psychology of Mathematics Education (IGPME) 40th conference taking place in Hungary in August 2016, one out of the four plenary speakers and one out of the five plenary panel speakers were from Israel. In the IGPME 37th conference in Germany in 2013, there were no less than 27 Israeli presenters (out of 327; 8%). In the same conference, there were eight presenters from Italy (2%) and five from France (1.5%), two countries that are geographically closer to Germany and their population is significantly larger than Israel's population.

Another strength related to the large faculty devoted to STEM education is the large number of doctoral students engaged in STEM education. Most of the proposals submitted to the conference were presented by doctoral students. Many of these students also work in the educational field, either in school or in teacher colleges, and can further foster opportunities for knowledge transfer between the education system and the academia by RPPs.

(c) **Additional Israeli research institutions in STEM education**

The third strength we have identified is the existence of institutions or organizations devoted to research and development in fields related to RPPs in STEM education. These include first-sector organizations (e.g., The Chief Scientist of the MOE[7]) and NGO nonprofit organizations (e.g., The Initiative for Applied Research in Education[8]).

Table 3.4 presents the ranking of each strength in the validation stage as well as selection of the most significant strength. As can be seen, the strength "multiple activities in STEM education" was ranked the highest with respect to its consideration as a strength, having been selected by 47.1% of the responders as the most significant strength. We note that the strength "large research community" was selected as the most significant strength by half of the responders from the

---

[7]See http://cms.education.gov.il/EducationCMS/Units/Scientist/Odot/afkideiMadan.htm.

[8]See http://yold.mpage.co.il/english/homepage.aspx. The Initiative for Applied Education Research was established in 2003 as a joint venture of the Israel Academy of Sciences and Humanities, the Israel MoE, and Yad Hanadiv. The initiative sets up expert committees and convenes symposia for researchers, education professionals, and decision-makers. It publishes reports of its work and makes them readily available to the public, which means that they have the potential to be read by teachers and principals and to be applied.

academia, but only by one-fifth of the responders from the other sectors. This can be explained by the fact that the academic participants are more familiar with these communities as well as with their activities. However, it can also point to a cultural gap, which will be discussed in the next section.

3.4.2 Weaknesses of RPPs in STEM Education in Israel

Our analysis revealed three main weaknesses regarding RPPs in STEM education in Israel: (a) a cultural gap between the academia and the educational field (schools and MoE), (b) academic incentives, and (c) regulations in the MoE. As can be seen, (a), (b), and (c) indicate that *not only a **cultural** gap exists between the academia and the education system, this gap is further fostered by incentives and regulations.* We mention that in addition to the three weaknesses mentioned here, four additional weaknesses were raised through the analysis of the conference data. However, they were found to be less meaningful during the validation stage and were therefore discarded or combined with more significant weaknesses.

(a) **Cultural Gap between the education system (schools and MoE) and the Academia**

All organizations, including nonprofit ones, have a culture which is expressed in the values, discourse, and behavior of their members, which in turn "contribute to the unique social and psychological environment of an organization."[9] A cultural gap exists when the values and behaviors of two (or more) organizations clash. Examination of the organizational culture is a useful tool for the analysis of organizations' successes and failures as well as for the communication styles among organizations.

In our case, the cultural gap between the academia and the education system was expressed in different ways and was evident from multiple sources. We address two expressions of this gap:

- The research needed for the MoE versus the research conducted in the academia;
- The spirit of the dialogue.

The research needed for the MoE versus the research done in the academia

This gap was expressed by several MoE representatives. As presented in Table 3.3, one of them (Asher) described two possible axes for the characterization of a research project: "close vs. open" and "fast vs. slow." He said that at the MoE, people are usually interested in "close and fast," meaning quick and closed answers to pressing questions. Researchers in academia, on the other hand, usually tend toward a "slow and open" orientation, meaning "give us several years and let us

---

[9]Business Dictionary. http://www.businessdictionary.com/.

explore open questions and then we will come up with some answers" (Asher, plenary response). As a result, many stakeholders in the MoE (according to Asher) conceive the idea of research as non-relevant, as something "theoretical," while in fact, it is empirical.

Time constraints and negotiation timelines were mentioned repeatedly in the pre-conference survey as challenges with which RPPs should deal. Importantly, time was often mentioned in relation to other resources and the trade-offs between them. For instance:

- "Lack of resources: time, money, personnel" (pre-survey response to question about limitations of RPPs).
- "Connection and coordination take time, effort, and compromises" (pre-survey response to question about limitations of RPPs).

Some also mentioned the incompatibilities more explicitly. For instance,

- "The agenda is not always the same agenda. The school wanted to use the research to promote its public relations, while for the academic institution it was more important to align with research ethics. At times, there is tension between partners that is caused by one partner feeling the other is taking advantage of it" (pre-survey response to question about limitations of RPPs).

The spirit of the dialogue

Another representative of the MoE said that the academia's attitude toward the "field" is arrogant in that people in the academia do not appreciate enough the activities that take place in the schools. Several comments in the post-conference survey explained this feeling by the kind of dialogue between the education system and the academia. For instance, when reflected on the conference, participants said:

- "It's sad, but I felt at times there was a dialogue of the deaf between the academia and the field. The attempt to rebuke the people of the MoE did not lead to any productive discourse" (post-conference survey).
- "It's a pity that as (people from the) academia, we weren't smart enough to really listen to what was said" (post-conference survey).
- "The MoE did not get sufficient voice. They should have been sitting in a panel of their own"[10] (post-conference survey).
- "The academia is detached from the field and provides ideas from a supremacy stance. First there is a need for trust and mutual learning before imposing innovative ideas" (pre-conference survey).

Trust is a key component of an organization culture. In the case of RPPs, without trust parents will not allow teachers and teachers will not allow researchers to do research at the schools (take notes, record videos, and observe classes).

---

[10]It should be mentioned that several people at the MoE were approached a few months before the conference and were asked to participate, yet their calendars did not permit it.

(b) **Academic culture and structure: Promotion, academic incentives, university reward system**

Our data revealed that communication barriers between the academia and the education system exist also from the academic perspective. One of the main reasons, talked about in the conference, was researchers' motives, which are the first and foremost tenure and promotion. In Israel, publications in Hebrew (the spoken language in most schools and academic institutions) are not viable as indicators of universities' productivity. Thus, the academic promotion process does not encourage researchers to publish in the main language of the country, leading to researchers publishing almost only in English, especially before they get tenure. This hinders dissemination and transfer of information from the academia to the education system. Thus, publishing incentives often determine a focus of research that is not well aligned with the pressing problems of the Israeli education system.

The CHE report, mentioned in the introduction, and discussed by the plenary speakers of the conference, addressed this issue[11]:

- "There is little incentive for scholars of education to disseminate their work beyond university walls, and thereby influence the society of which they are a part" (p. 8).
- "The impulse to distance oneself from the core problems of education sometimes comes from the highest levels of the university"(p. 10).
- "What attracts faculty attention is the university reward system, where the emphasis is on publishing in English-language journals" (p. 12).
- "At present the main audience for the research produced by Israel's education scholars are the professors publishing in the same journals. There are few opportunities for practitioners to learn about the kind of research that might improve their practice" (p. 14).

Another strong set of incentives is national and international funds. As presented in Stein's keynote in our conference, many RPPs in the USA are supported by national foundations such as the NSF and IES grants. For example, Stein presented projects such as COMET and "Scaling up Mathematics,"[12] funded by the NSF which included a strong implementation and dissemination aspect of research-based programs for mathematical teaching and learning. The ISF, on the other hand, does not promote any RPPs; on the contrary, it stresses that it only funds basic research.

Motives of researchers and promotion criteria of course do not exist in a vacuum. They are a result of the way in which the CHE rewards Israeli universities, which is according to the number and status of publications in international journals, number of grants, number of doctoral students, etc. This incentive structure contrasts sharply with the CHE report just quoted that criticized education departments for

---

[11]The authors of this paper express some reservations about some of the more absolutist statements of this report.

[12]See https://www.nsf.gov/awardsearch/showAward?AWD_ID=0228343&HistoricalAwards=false.

engaging in research activities that align precisely with the incentive structure for which it is responsible.

(c) Regulations of the MoE

The third weakness we identified related to organizational structure and regulations of the MoE. Specifically, from the organizational structure perspective, we address the teaching position; from the regulation perspective, we address the approval process of research proposal submissions to the office of the Chief Scientist of MoE.

The teaching position in Israel

Evidence from the participants' stories, as well as follow-up conversations with major figures in the MoE reveal that the uniform teaching position in Israel (including uniform wages), its rigid structure, and lack of benefits teacher gain from participation in RPPs block teachers from actively participating in research projects. Incompatible incentives and motives were mentioned repeatedly in the data. For example:

- "Teachers do not have incentives to collaborate and therefore it is difficult to find places to carry out the research. Also, the teachers who help and support to carry out the research are not committed: The position structure, payment, professional training, overwork" (challenges mentioned in a project abstracts).
- "Institutional structures that would fit and support the research program and its principles, including changes in teachers' wages agreements, compensating teacher teachers, etc." (suggestions for a change mentioned in a project abstract).

In many cases, research studies in STEM education aim at developing new tools, teaching methods, etc., that require a teacher professional development program. However, as the following quotes indicate, in the current structure of the teaching position, teachers are not committed to such professional development programs, since they are required to participate in professional development programs that take place in their schools.

- "Difficulties in recruiting teachers to the professional training; lack of full commitment to participate in the program; time constraints of the professional training; teachers' unwillingness to continue working from home" (challenges mentioned in a project abstract).
- "Difficulties in recruiting teachers: voluntarily the willingness is almost zero and in teacher professional training the teacher arrive very tired and many of them do not arrive at all" (challenges mentioned in a project abstract).
- "The teacher leaders contend with school principals, superintendents, MoE demands, instructions from the subject supervision unity, etc., and they are influenced by decisions like [...] school professional development within Oz Latmura [a new reform in the structure of teacher position] which make participation in out-of-school PLCs [professional learning community] difficult" (challenges mentioned in a project abstract).

It is important to note the use of the term recruiting teachers, mentioned in two of these examples, which indicates that teachers are conceived as "participants" or even "subjects" rather than equal collaborators. This perspective can be explained in two ways:

(a) The prominence of the linear model in the educational research community, which treats teachers as receivers of research-manufactured knowledge, not as the creators of it.
(b) Limited resources, which pose difficulties to attract teachers to actively participate in RPPs. Several project abstracts explicitly declared this message, e.g., "The main obstacle is funding and the secondary obstacle is *recruiting* teachers and students for research and implementation."

Regulations regarding research approval in the MoE[13]

The Chief Scientist of the MoE holds the responsibility to design the Ministry's research policy, including setting criteria for allocating research funds, setting priorities for research in different domains of education, and creating research-based collaborations with government, international, and non-government institutions as well as integrating scientific knowledge that can serve policy makers' decision-making processes in the MoE.

Another role of the Chief Scientist of the MoE is to serve as an IRB—Institutional Review Board—so that every research conducted in the education system in Israel should get the approval of the Chief Scientist. The purpose is to verify that pupils' and teachers' rights are respected and ethical research norms are adhered to.[14] However, the limited resources allocated to the Chief Scientist produce a long queue of research proposals submitted for this authority's approval. These were evident in the conference participants' complaints about significant delays in the approval process and, consequently, inability to perform the research as planned. This is especially important when graduate students need to be committed to their university schedule for performing their research. The complaints expressed in the workshop are supported by a recent report, published by the Israeli Knesset's Center for Research and Information,[15] about the role and status of the Chief Scientists in the governmental ministries in Israel (Goldschmidt 2016). According to this report, the Chief Scientist office at the MoE is one of the under budgeted Chief Ministry offices relative to the budget of the ministry, specifically, the budget allocated for the Chief Scientist office at the MoE equals 0.02% of the total budget of the MoE (Goldschmidt 2016, p. 9).

To ease the Chief Scientist's administrative overload, several rules have been updated recently. According to these rules, two types of research have been exempt

---

[13]See: http://cms.education.gov.il/EducationCMS/Units/Scientist/Odot/afkideiMadan.htm.

[14]The submission guidelines for a request for data collections in schools are presented here (in Hebrew):   http://cms.education.gov.il/EducationCMS/Applications/Mankal/EtsMedorim/3/3-9/HoraotKeva/K-2015-9-2-3-9-4.htm.

[15]The Knesset is Israel's parliament.

from approval of the Chief Scientist: (a) research works that do not focus on students and where data collection encompasses less than 15 educational organizations and (b) a case study in one educational organization that does not focus on students and was initiated by its principal.[16] In addition, the principal/manager of the educational organization in which the research work takes place should approve the data collection and all research works should be approved by the research ethics committee of the researchers' university.[17]

In spite of these changes, these two exemptions do not pertain to most of the studies conducted by academic researchers in the education system, since those usually do include students. References to difficulties having to do with MoE regulations appeared in some of the project abstracts and were even more prevalent in informal conversations with participants. Quotes from project abstracts include:

- Difficulties in getting approval of the chief scientist unit—even though we sent all the files and responded quickly to the questions raised, we got permission only after a long time and even then it was only partial (project abstract).

For extreme illustration, we present one of the participants' stories, who reported on her project that included development of curriculum materials. She wrote:

- As the developers of the textbooks, we were not allowed to enter and observe classrooms (pre-conference survey).

Furthermore, as mentioned, according to the regulations of the Chief Scientist of the MoE, each research that takes place in the school requires the principal's approval. Currently, school principals do not have much to gain neither from a research-oriented incentive nor practical-oriented incentive. They thus do not have incentives for granting such permissions. Here is a demonstrative quote, from a project abstract describing the challenges of the project:

- "Difficulties in getting the principals' approval, despite the fact that a permission was obtained from the office of the Chief scientists. When I approached the school principals, most of them did not respond at all and only a few agreed [to allow me to collect data in the school]" (project abstracts).

Given the linear model that we saw prevalent in most projects, where schools served as a "field" for enacting a certain pre-planned study, such a mode of action on the principals' side makes sense. When positioned as "subjects" rather than as collaborators, principals have little incentives for joining research initiatives. We believe (and propose later in the chapter) that if principals would be approached as collaborators, and see tangible benefits from their participation in such projects, their approval would be received more readily.

---

[16]See CEO notice 3.9-4 from the Ministry of Education at http://cms.education.gov.il/ EducationCMS/Applications/Mankal/EtsMedorim/3/3-9/HoraotKeva/K-2015-9-2-3-9-4.htm.

[17]See, for example, the Technion's committees: http://manlam.net.technion.ac.il/en/ethics_committees/.

Table 3.4 shows that all responders were familiar with the weakness "cultural gap between the academia and the education system." This level of familiarity was not reflected with respect to any of the other factors in this section as well as among the other factors presented in the other sections of the SWOT analysis. Furthermore, it got the highest ranking with respect to its consideration as a weakness (only the threat "priorities change" received similar ranking with respect to its category as being a threat). In addition, it got the highest selection percentages as the most significant weakness (this weakness was selected as the most significant weakness in the validation stage by 40.6% of the responders, while the other options were selected significantly lower as the most meaningful weakness).

3.4.3 Opportunities of RPPs in STEM Education in Israel

In this section, we review two opportunities that arose from the data. These are opportunities that both the academia and the education system could take advantage of, in order to promote RPPs.

(a) **Cross-sectoral—beyond the public sector—awareness to STEM education in Israel: the industry, nonprofits, and philanthropy**

In recent years, the Israeli field has seen a significant increase in the awareness of the importance of STEM education. This awareness has stemmed from several reasons, among them the need for professional human resources, especially engineers, required by the hi-tech sector in Israel, which has a crucial role in Israel's economic and development growth.

One of the industry's representatives in the conference described a collaboration around raising the number of students taking mathematics at a 5-unit level as an example of cross-sectoral collaborations aimed at one broad national problem, which none of the partners have the ability to solve alone.

- Our model says that by bringing in more partners with diverse abilities into the activity, we in fact more than double, we quadruple the shared value (of our investment). So, beyond the business value for the organization, there is a social value and this value grows more and more (Abrahams et al. 2015).

The investment efforts and attention are all geared to the effort of preserving Israel as the "start-up nation" (Sensor and Singer 2009).

(b) **The CHE evaluation report as an opportunity**

Though the CHE evaluation report was mentioned during the conference mainly as an indicator of weaknesses, we believe its current prominence in the Israeli academic discourse serves as an important opportunity.

As mentioned in the introduction, during 2012–2014, the seven departments of education in all Israeli universities were evaluated by an international committee of the Israeli Council of Higher Education (CHE). With respect to research, the report indicates that:

- Investigating issues like the Bagrut [matriculation exams], or subjects unique to Israeli curriculum like Mikrah [Bible], or the challenges of an educational system divided four ways into secular, religious-Zionist, Arab and Haredi [Ultra-orthodox] systems, each with its own goals, ideologies, and attitudes toward the state, are seen as less worthy than appealing to editors of international journals. It is a bitter irony that the issues that make Israel's educational system unique are the ones least studied by its scholars (CHE 2015, p. 12).

Though this report was general and described an overall picture (that is, not only on STEM education), it is reasonable to assume that this criticism is also correct, at least partially, with respect to STEM education (as the "academic culture and structure" weakness described above indicates).

In addition to raising awareness and debate, the report has encouraged schools of education to re-evaluate their programs and their connection to the education system.[18] Through its public accessibility, the report can serve as a point of reference for all parties involved in the dialogue to improve RPPs. In that sense, it has turned "corridor talk" into a viable artifact around which future efforts for improvement can be constructed.

As given in Table 3.4, the opportunity "cross-sectoral awareness" was considered by the validating respondents as the major opportunity that should be exploited in order to promote RPPs. Twenty-five (25) respondents out of 29 chose it as an the most important opportunity, either with relation to scores in international tests, or with relation to the demands of the high-tech community for STEM-educated working force. The opportunities enfolded in the CHE were less clear to many of the respondents, and only four chose it as the most significant opportunity (though it did receive a moderate score of agreement—2.0, in relation to being identified as an opportunity). This seems reasonable as the CHE report is still mostly known in relatively small circles.

3.4.4   Threats of RPPs in STEM Education in Israel

In this section, we review two threats that have been discussed by conference participants, which belong to factors outside the academia and the education system and block the promotion of RPPs: Priorities change due to the frequent

---

[18]In fact, our faculty treated this evaluation process as an opportunity to upgrade its status from a department to a faculty—with an equal status to all other Technion's academic units. In addition, at our faculty, a major reform in the structure of basic undergraduate studies in education has been started recently based on the feedback we received in the evaluation report. Specifically, we modified the basic mandatory courses which were mostly in the domain of theory and psychology, to courses that are much more ingrained in school life and in current theories of learning. We hope that this change will foster also RPPs.

appointments of new Ministers of Education and the fact that the responsibility of the education system is consistently transferred from the state to private (for-profit and nonprofit) organizations.

(a) Priorities change

One of the major threats is the political instability and the penetrability of the MoE to political influence. In the past decade, six ministers of education from five political parties have changed in Israel according to shifts in political power. Each minister has brought with him/her a new agenda and a major shift in priorities. Since the entry of Nafatli Bennet in 2015 to the MoE office, the focus has shifted to promoting excellence in mathematics and sciences. Each one of these goals, in itself, is very important and aligns well with the goals of STEM education. However, the rapid shifts in priorities of the MoE make it difficult for agents both in the MoE and out of it to plan for any program in general and for any research in STEM education in particular—and specifically, RPPs—that is beyond a few months. Not surprisingly, 83.9% selected this threat as the most significant threat from the two proposed threats.

(b) Responsibly of the education system is transferred from the state to for-profit and nonprofit NGOs

As participants in the conference mentioned, the involvement of the state in funding of research and development in the field of STEM education has significantly decreased in the past few years. The void left by the state has attracted many agents of the second (industry) and third (NGOs) sectors, who, as reported in the conference, engage in and support some very influential projects in general and in STEM education in particular.

Unlike public sector funding, the second- and third-sector funding usually does not go out through public calls but rather through private initiatives that do not include funding for research. This shift is further supported by The Mandatory Tenders Regulations No. 5753-1993.[19] As in many places around the world, Israel's Mandatory Tenders Regulations[20] require public organizations and government to

---

[19]This regulation was not discussed broadly in the workshop. However, once it was mentioned by one of the participants, we delved into its details and realized that it might have a crucial role as a factor that prevents the execution of RPPs related to activities carried out in STEM education in Israel.

[20]Source: https://www.mr.gov.il/Information/Training%20materials/Mandatory%20Tenders%20 Regulations.pdf, http://www.economy.gov.il/English/InternationalAffairs/IndustrialCooperation Authority/Pages/MandatoryTendersRegulations.aspx.

Specifically, as it turns out, though quality is highly sought, the law indicates that price plays a role:

(2) Following the opening of the tender box, the tender committee shall, according to the Mandatory Tenders Regulations, 5753-1993 Complete up-to-date version 27, determine the final group of bidders and the quality score for each bidder.

(3) The price bids shall be opened only after the tender committee has determined the quality score. Following the opening of the price bids, the tender committee shall determine a final score

publish tenders prior to the purchase of some merchandise and for recruitment of human resources from a certain level of cost. As it turns out, while the rationale behind the law is fully understandable as being aimed at providing equal opportunities to all providers, it turns out that it blocks the inclusion of the research component in general and of RPPs in particular. This was evident from the talk of one of the conference participants who had served in a primary role in the MoE and felt these regulations had seriously constrained her actions in the domain of RPPs.

In practice, when calls are published by the MoE for the development of curriculum materials (such as textbooks, software tools, and teacher professional programs), the universities submit proposals that include research, whereas for-profit and other nonprofit NGOs submit cheaper proposals that do not include research. In many cases, budget constraints force the MoE, by this regulation, to accept the cheaper proposals. Accordingly, research institutions and universities lose the bid to other for-profit and nonprofit organizations and the education system loses an opportunity to learn from the development process through research. It is important to mention that the Knesset's report mentioned above (Goldschmidt 2016) also asserts that the regulation of merchandise purchases does not fit the purchase of research "products" (the quotation mark appears in the original, p. 13).

The result of this process, as pointed out by one of the academic scholars in the conference's audience, is that "the responsibility for the educational agenda moves to private (either second sector or third sector) hands" and the second and third sectors are perceived as key factors in the education system in Israel. This forms a threat for several reasons, both on the state of Israel level and on the RPPs level discussed in this chapter:

(I)   Researchers, who insist on carrying out research in schools, approach private funds that eventually determine the national agenda.

(II) Most industrial funds and private foundations do not advertise public calls. Equitable opportunities for researchers in the academia and practitioners in the education system to win funding are thus greatly reduced.

(III) As mentioned by one of the representatives of NGOs in the panel, private NGOs rarely have the resources and means to perform rigorous peer-review processes to assess the quality of proposals, so it is not clear that the best programs are eventually funded.

(IV) The same representative also mentioned the chaotic situation that exists as a result of many uncoordinated projects managed by these NGOs.

(V)   Finally, and most important, this shift transfers the responsibility and control of the education system from the MoE to NGOs which sometimes have different interests and leave the state in a meager situation, unable to promote policy that is most beneficial for the general population and particularly for disadvantaged sectors.

(Footnote 20 continued)

for the bids that weights the quality score determined as provided in paragraph (2) with the score based on the price.

As Table 3.4 shows, while "priorities change" received the highest agreement scores for being a threat, the factor we named "responsibility is transferred from the state to private organizations" got the lowest ranking, meaning it was identified the least as a threat by validating respondents. These rankings reflect a conflict: on the one hand, clear leadership in the MoE is desired and the frequent changes in the Minister of Education office matter; on the other hand, the fact that the government transfers the responsibility of education to third-sector organizations is not conceived as critical.

3.5   First Steps and Recommendations for Bridging the Cultural Gap Between Academia and the Education System

SWOT analysis provides organizations a useful tool for strategic analysis and planning. It guides an action in which the Strengths are leveraged to overcome the Weaknesses, in a way that utilizes the Opportunities existing outside the organization, while accepting the external Threats and continuing to operate successfully despite them. In this section, we propose several recommendations that can be considered to promote RPPs in STEM education in Israel.

In order to promote successful RPPs in STEM education in Israel, we address organizational aspects of the academia and the MoE in order to close the cultural gaps existing between the education system and the academia (a weakness). This can be done by using the rich infrastructure of STEM education practice and research existing in Israel (a strength) and utilizing the increased awareness and importance attributed to STEM education in Israel (an opportunity). Since the weakness components of the SWOT analysis were significantly more evident than the other components of the SWOT analysis, we hope that our proposed recommendations may help exploit how the identified strengths and the opportunities can be utilized in order to close the gap.

We suggest that the alternative model offered by Stein and Coburn (2010) (see Fig. 3.1) may guide us in the application of these recommendations. As mentioned above, to reflect the essence of the alternative model, we call it the "dual model."

The following recommendations for the establishment of the *Dual Model in STEM education in Israel* were validated mainly in two ways—before the conference and through the post-conference validation stage, as described below:

- The pre-questionnaire: In two questions in this questionnaire, the participants were asked to choose three items from a list of nine items that they conceive as the most meaningful channels for information flow from the education system to the academia and from the academia to the education system. The nine items were: books and learning materials, graduate students, research projects, academic staff, adjunct lecturers, teacher preparation program, teacher professional development programs, visit in schools, and others.

With respect to the channels from the *education system to the academia*, the following four channels were indicated as the most meaningful ($n = 58$):

(a) Visits in the schools (39 answers, 67.2%).
(b) Teacher professional development programs (30 answers, 51.7%).
(c) Graduate students (22 answers, 37.9%).
(d) Research projects (22 answers, 37.9%).

With respect to the channels from the *academia to the education system*, the following four channels were indicated as the most meaningful ($n = 58$):

(a) Teacher professional development programs (45 answers, 77.6%).
(b) Teacher preparation programs (41 answers, 70.7%).
(c) Graduate students (27 answers, 46.6%).
(d) Books and learning materials (27 answers, 46.6%).

As can be seen, the channel indicated as the most meaningful for bidirectional information transfer (both from the education system to the academia and vice versa) was *teacher professional development programs*. Clearly, as an information channel that is conceived meaningful in both directions, it has the potential to form a meaningful infrastructure for the Dual Model for RPPs in STEM education in Israel. It also has the potential to close the cultural gap existing between the MoE and the academia.

- The validation survey: In this survey, we proposed recommendations for the promotion of STEM education in Israel, and for each of them, the participants were asked to rank its appropriateness as a possible action for the promotion of RPPs in Israel (Table 3.5). As can be seen, the recommendation "put more emphasis in the academia on research that incorporates implementation in the educational field" was ranked high absolutely and relatively to the other proposed recommendations. In addition, 25 participants out of 37 ranked it as "extremely high" as an appropriate action for the promotion of RPPs in STEM education in Israel, and no one ranked it as irrelevant at all. *All these indicators imply that the responsibility for the promotion of RPPs largely depends on the academia.* This action indeed makes sense since the research part of the RPPs should push the implementation part and the academia has this expertise.

In light of the cultural gap existing between the academia and the education system (as extensively described in the Weaknesses section), we propose two recommendations for bridging this gap: emphasizing teacher professional development and communication channels to maintain the emerging dialogue in a systematic way.

(a) **Teacher professional development as an area for RPPs**

As mentioned in the Weaknesses section, lack of teacher participation in research is one of the main barriers that block the flourishing of RPPs. Accordingly, ways to foster teacher participation in research and in the discourse about the desired kind of

Table 3.5 Validation of suggested recommendations for the promotion of RPPs

Recommendations N = 41	Put more emphasis in the academia on research that incorporates implementation in the educational field	Incentives for teachers and principals to participate in research	Establishment of an "entity" whose role is to promote RPPs, maintain the discourse and bridge the cultural gap[a]	Use "Ofek Chadsh" (New Horizon[b]) platform to foster professional development programs around research	Use "Academia-Kita" (Academia-Classroom) platform for the promotion of RPPs
N	37	39	37	36	28
Average (ranking out of 3)	2.65	2.33	2.32	2.11	2.10
Not at all-0	–	1	1	2	1
Low-1	1	5	5	3	6
High-2	11	13	12	20	10
Extremely high-3	25	20	19	11	11
Do not know/did not answer	4	2	4	5	13

[a]This entity is expressed in our proposed actions as communication channels between the education system and the academia

[b]New Horizon (Ofek Hadash) is an educational and professional reform in elementary and junior high school education. Its implementation began in 2008

RPPs should be sought after. Further, it is proposed that the inclusion of teachers in research projects might be a good solution for research sustainability.

We propose that RPPs can be fostered by the promotion of the "teachers as researchers" conception (Sahlberg 2011, pp. 83–86). This can be done in several ways, closely related to the structure of the teaching position (see Weaknesses). Here are several actions in this direction:

- *Orientation*: Today, teachers often seek masters and Ph.D. certificates as a way *out* of schools (e.g., to get instruction jobs in teacher colleges). Career incentives should make masters and Ph.D. studies beneficial for staying *in* school and carrying out or implementing research findings within it. This is a timely era to promote this direction of "teachers as researchers," especially in STEM education in Israel, since more and more highly educated scientists and engineers join the education system as a second career, due to two main reasons: first, the strong commitment of Israeli citizens to contribute to the society and, second, the hi-tech job market in which (a) scientists and engineers who do not wish to be promoted to managerial roles find themselves without a job, and (b) young scientists and engineers are sometimes preferred over experienced ones. (e.g., Hazzan and Ragonis 2014; Gero and Hazzan 2016).
- *Action research* (Lewin 1946): Intensively include a research-oriented thinking and practice in teacher preparation programs, as well as in professional development programs of teachers, principals, and inspectors. This recommendation supports Boyd et al.'s (2011) suggestions that "policies aimed at improving school administration may be effective at reducing teacher turnover. [...] Current reforms aim to recruit high-potential leaders, provide apprenticeship experiences for prospective leaders, and to provide supports for principals while in the job. Improving administrative support in high-turnover schools in particular may require both more effective leaders, overall, and incentives (not necessarily monetary) so that administrative positions in these schools become more appealing" (p. 329).
- *Structure of the teaching position*: A new design of the teaching position should support research opportunities, e.g., by a 1–2 days free of teaching as part of the position, and the establishment of research communities which include teachers and researchers. This structure should not be applied for all teachers in the first stage, but rather to a selective group of teachers who are attracted by and committed to this opportunity and can guide the education system in this direction. The research community needed for the promotion of this action can be established in professional development programs[21] integrated through education reforms that have been launched recently by the MoE ("Ofek

---

[21]As mentioned in the Weaknesses section, the current structure of the teaching position does not encourage teachers to commit to professional development programs. However, at the same time, professional development programs were identified as a meaningful communication channels between the academia and the education system. Therefore, it is proposed to redesign them in a meaningful way for the promotion of RPPs within the redesign of the teaching position.

Chadash," that is, New Horizon[22]) and the Academia-Kita—Academia-Classroom[23] framework introduced recently as part of teacher preparation programs. These programs deliver the message that the education system wishes to collaborate with the academia. However, they do so according to the linear model. They encourage professional development to be led by universities and colleges, but do not promote research through these professional development programs. We propose that RPPs may turn these opportunities into important factors in the creation of the Dual Model for RPPs.

(b) Communication channels between the education system and the academia

Finally, in order to maintain the discourse between the education system and the academia to bridge the cultural gap consistently, we propose to strengthen the existing communication channels that can be naturally integrated in the activities of one of the involved body—the MoE (e.g., in the office of the Chief Scientist), the academia and second- and third-sector organizations. For example, doctoral students who often work in both systems—the education and the academia—can serve as bidirectional avenues of communication. Though only 40% of the responders indicated them as import communicational channels, we believe that their potential to serve as bidirectional avenues of communication has not been sufficiently exploited so far. With the right support, and maybe some institutional measures such as program requirements, their contribution for bridging the cultural gap can be enhanced.

In general, communication channels between the academia and educational practice should reflect the model shift from "providing research services" to a model of partnership. They should deliver a clear message: Research is important not only for academic people but also, and maybe even more important, for school principals and teachers, who can improve their practice by active participation in research projects. This suggestion is supported by Cole and Knowles (1993), who argue that "new forms of partnership research are based on fundamental assumptions about the importance of mutuality in purpose, interpretation, and reporting, and about the potency of multiple perspectives. Also implicit in this model is the understanding that each partner in the inquiry process contributes particular and

---

[22]The New Horizon reform promotes teacher professionalism by defining levels of expertise. One of the highest levels (Level 8) is Teacher-Researchers. Teachers who are promoted to this rank are expected to integrate research in their work in order to improve their teaching processes. This reform provides suitable opportunities for the promotion of RPPs since: a) teachers who are promoted to this rank get salary increase and b) professional development programs are funded to let teachers gain the needed research skills.

[23]The Academia-Classroom reform attempts to foster the connections between the teacher training programs, which usually take place in the academia, and the school system. This is done by teaching several courses of the teacher preparation programs in the schools and the integration of the pre-service teachers in actual teaching processes in the classrooms. This structure naturally provides many opportunities for RPPs since the academic people, who teach the courses, visit the schools as part of their teaching, and interact on a weekly with the new teachers and their mentors.

important expertise, and that the relationship between the classroom teacher and the university researcher, for example, is multifaceted and not powerfully hierarchical" (p. 478).

3.6 Conclusion

This chapter analyzes RPPs in STEM education in Israel by exploring the perceptions of practitioners from the education system and from the academia with respect to RPPs they are currently involved in, as well as how such RPPs can be enhanced. The discourse about the topic is not carried out in a vacuum; it is deeply rooted in the general agreement, validated in our work, that such RPPs may promote Israel's achievements in the STEM subjects and contribute to Israel's competitive advantages in these fields.

The findings section of this chapter presents SWOT analysis of the current situation of RPPs in Israel. Table 3.6 summarizes the factors identified in each component of the SWOT analysis. In addition, we suggest recommendations that can be carried out in order to promote RPPs in STEM education in Israel. Change processes are long (sometimes several decades) and require from all sectors a conceptual change. However, as other countries tell us (e.g., Finland—see Sahlberg 2011) it is doable.

The current work also has several limitations. First, due to the forms in which data were collected, it does not necessarily represent all views in the academia, education system, or general Israeli public. Second, it is tainted by our own personal beliefs, which view RPPs and involvement of researchers in educational practice as valuable and important. It is because of this belief that we advanced the

Table 3.6 High-level SWOT analysis of RPPs in STEM education in Israel

Strengths	Weaknesses
1. Multiple activities in STEM education	1. Cultural gap between the education system (schools and MoE) and the Academia
2. Large research communities in the different STEM subjects in higher education units	2. Academic culture and structure: promotion, academic incentives, university reward system
3. In addition to the academic units, several other research in education institutions exist	3. Regulations in the MoE
	1. The teaching position
	2. Chief Scientist of MoE
Opportunities	**Threats**
1. The Council of Higher Education (CHE) evaluation report	1. Priorities change
2. Cross-sectoral—beyond the public sector —awareness to STEM education in Israel: the industry, nonprofits, philanthropy	2. Responsibly of the education system is transferred from the state to for-profit and nonprofit NGOs

conference and this study around it, and it is this belief that currently leads us to continue pursuing ways to promote RPPs in STEM education.

This chapter is not only a report of data and its analysis; it has also been serving, through its writing process, as a continuation of the dialogue between the academia and the MoE. We are currently seeking ways, together with the Acting Chief Scientist and with figures from the third sector, to propel this dialogue forward in an ever-growing attempt to start closing the cultural gap between the academia and the educational field.

We hope to continue our study in a way that expands the research population and scope, engages more teachers and school principals in the discussion, evaluates new initiatives that will be launched as a result of this multi-sectoral discourse, and finally considers the participation of additional bodies in Israel in the discussion, such as academic colleges and teacher colleges.

Acknowledgements This work was partially supported by Heyd-Metzuyanim's Spencer Small Grant number 201500080.

References

Abrahams, B., Waksman, M., Hazzan, O. and Lis-Hacohen, R. (2015). The (CS) 2V—Cross sectorial collaborative shared value—strategy. Unpublished position chapter: http://edu.technion.ac.il/Faculty/OritH/HomePage/CSCSV_Strategy_March2016.pdf.

Avargil, S., Herscovitz, P., & Dori, Y. J. (2013). Challenges in the transition to large-scale reform in chemical education. *Thinking Skills and Creativity, 10,* 189–207.

Barney, J. (1995). Looking inside for competitive advantage. *The Academy of Management Executive, 4,* 49–61.

Boyd, D., Grossman, P., Ing, M., Lankford, H., Loeb, S., & Wyckoff, J. (2011). The influence of school administrators on teacher retention decisions. *American Educational Research Journal, 48*(2), 303–333. doi:10.3102/0002831210380788.

Coburn, C. E., & Stein, M. K. (2010). *Research and practice in education: Building alliances, bridging the divide.* Rowman & Littlefield Publishers.

CHE (Council of Higher Education). (2015). *Committee for the Evaluation of Education and Science Education Study Programs General Report* http://che.org.il/wp-content/uploads/2016/04/Education-and-Science-Teaching-QA-Commitee-General-Report.pdf.

Coburn, C. E., & Penuel, W. R. (2016). Research-Practice Partnerships in education: Outcomes, dynamics, and open Questions. *Educational Researcher, 45*(1), 48–54.

Cole, A. L., & Knowles, G. (1993). Teacher development partnership research: A focus on methods and issues. *American Educational Research Journal, 30*(3), 473–495.

Even-Zahav, A. (2016). *Strategic analysis of educational systems: Risk management of STEM (science, technology, engineering and mathematics) education in Israel,* Ph.D. dissertation under the supervision of Orit Hazzan, Technion.

Gero, A., & Hazzan, O. (2016). Training scientists and engineers as science and engineering teachers: The motivational factors of enrolees in the Views programme. *World Transactions on Engineering and Technology Education, 14*(3), 374–379.

Goldschmidt, R. (2016). *The Role and Status of the Chief Scientists in the Governmental Ministries in Israel,* The Israeli Knesset's Center for Research and Information, http://fs.knesset.gov.il//20/Committees/20_cs_bg_343503.pdf (Hebrew).

Hazzan, O., & Lis-Hacohen, R. (2016). *The MERge model for business development: The Amalgamation of Management, Education and Research*, SpringerBriefs in Business. http://www.springer.com/us/book/9783319302249 (120 pp.).

Hazzan, O., & Ragonis, N. (2014). STEM teaching as an additional profession for scientists and engineers: The case of Computer Science education. In *Proceedings of SIGCSE 2014—The 45th ACM Technical Symposium on Computer Science Education*, Atlanta, GA, USA, pp. 181–186.

Kania, J., & Kramer, M. (2011, Winter). *Collective impact, Stanford Social Innovation Review*, http://ssir.org/articles/entry/collective_impact.

Kon, F., Cukier, D., Melo, C., Hazzan, O., & Yuklea, H. (2014). *A conceptual framework for software startup ecosystems: The case of Israel*. Technical Report RT-MAC-2015-01, June 2015. http://www.ime.usp.br/~kon/chapters/SoftwareStartupsConceputalFramework-TR.pdf.

Lewin, K. (1946). Action research and minority problems. *Journal of Social Issues, 2*(4), 34–46.

Lowy, A., & Hood, P. (2010). *The power of the 2 × 2 matrix: Using 2 × 2 thinking to solve business problems and make better decisions*. San Francisco: Jossey-Bass.

Rego, G., & Nunes, R. (2010). Hospital foundation: A SWOT analysis. *IBusiness, 2*, 210–217.

Sabbaghi, A., & Vaidyanathan, G. (2004). SWOT analysis and theory of constraint in information technology projects. *Information Systems Education Journal, 2*(23), 1–19.

Sahlberg, P. (2011). *Finnish lessons*. Teachers College Press, Teachers College, Columbia University, New York,USA.

Sensor, D., & Singer, S. (2009). *Start-up nation: The story of Israel's economic miracle*, Twelve, New York, USA..

Stein, M. K., & Coburn, C. E. (2010). Reframing the problem of research and practice. In C. E. Coburn & M. K. Stein (Eds.), *Research and practice in education: Building alliances, bridging the divide* (pp. 1–13). Lanham, MD: Rowman & Littlefield.

Penuel, W. R., Allen, A., Coburn, C. E., & Farrell, C. (2015). Conceptualizing research–practice partnerships as joint work at boundaries. *Journal of Education for Students Placed at Risk (JESPAR), 20*(1–2), 182–197. doi:10.1080/10824669.2014.988334.

Zohar, A. (In Press). Wide scale implementation through capacity building of senior leaders: The case of teaching thinking in Israeli schools. In D. Hung., S. S. Lee., Y. Toh., L.K. Wu., & A. Jamaludin (Eds.), *Innovations in Educational change—Cultivating Ecologies for Schools*. Singapore: Springer.

Zohar, A., & Cohen, A. (2016). Large scale implementation of higher order thinking (HOT) in civic education: The interplay of policy, politics, pedagogical leadership and detailed pedagogical planning. *Thinking Skills and Creativity, 21*, 85–96.

Epilogue
SWOT analysis of STEM Education: How can the World Benefit?

Why is this Brief important? We outline seven main reasons:

1. The Brief addresses different kinds of educational organizations whose focus is STEM education.
2. The Brief analyzes different kinds of educational organizations from a managerial perspective in general and SWOT analysis in particular. Such examination, which is very common in the case of for-profit organizations, sheds new light on educational organizations.
3. Educational organizations should be treated as any other types of organizations. If SWOT analysis benefits and contributes to other organizations, educational organizations should examine how they can also benefit from such an analysis.
4. Data in general and data about educational organizations in particular should be available and transparent; decision can be made based on this data. SWOT analysis is only one way to use this data; other ways should be revealed and explored as well.
5. Educational organizations such as schools, colleges, and university departments often need to undergo major changes in line with current reforms in education. SWOT analysis can direct them in such processes.
6. Collaboration between different types of organizations is crucial. Chapter 3 highlights the challenges of such collaborations, i.e., the collaboration between the education system and university researchers. SWOT analysis can help such organizations recognize common targets and interests, as well as opportunities, to overcome gaps and miscommunications in order to better exploit their resources and strengths.
7. Educational organizations cannot avoid competition. SWOT analysis may help them identify their strengths and direct them how to use these strengths both to overcome their weaknesses and in promoting their unique characteristics and achievements.

© The Author(s) 2018
O. Hazzan et al., *Application of Management Theories for STEM Education*,
SpringerBriefs in Education, https://doi.org/10.1007/978-3-319-68950-0

Printed in the United States
By Bookmasters